Refriger

ALSO BY CARROLL GANTZ
AND FROM MCFARLAND

Founders of American Industrial Design (2014)

The Vacuum Cleaner: A History (2012)

*The Industrialization of Design:
A History from the Steam Age to Today* (2011)

Refrigeration

A History

CARROLL GANTZ

McFarland & Company, Inc., Publishers
Jefferson, North Carolina

Library of Congress Cataloguing-in-Publication Data

Gantz, Carroll, 1931–
Refrigeration : a history / Carroll Gantz.
p. cm.
Includes bibliographical references and index.

ISBN 978-0-7864-7687-9 (softcover : acid free paper) ∞
ISBN 978-1-4766-1969-9 (ebook)

1. Refrigeration and refrigerating machinery—History. I. Title.

TP492.G29 2015
621.5'6—dc23 2015023132

British Library cataloguing data are available

Printed in the United States of America

McFarland & Company, Inc., Publishers
Box 611, Jefferson, North Carolina 28640
www.mcfarlandpub.com

This book is dedicated to the many scientists who discovered the principles of refrigeration, the practical engineers and entrepreneurs who designed and built the early machines to produce artificial ice and refrigeration, and the designers who transformed unattractive mechanical products into fashionable and user-friendly modern home appliances.

Table of Contents

Acknowledgments

I would like to acknowledge a number of people who have made this book possible. The history of refrigeration is scattered in a number of corporate and history websites, often in fragmented, rather than complete form. The only comprehensive book on the subject I found was a 1994 book published by the American Society of Heating, Refrigerating, and Air Conditioning Engineers (ASHRAE) with the title *Heat and Cold*. It covers both heating and refrigeration history in depth. Its authors are quite knowledgeable about the subject, and I am indebted to them for considerable references to their text and illustrations in this book. For those who would like to know more about heating and cooling history, I highly recommend it. However, *Heat and Cold* ends in about 1930, before household refrigerators became common; this book continues the story of refrigeration to the present day, and includes the complete design history of common household refrigerators.

I was fortunate to be helped by my very good friend and design historian colleague, Victoria Matranga, H/IDSA, who provided images of 1950s refrigerators and the unusual "icyball." Victoria is design projects coordinator and historian for the International Housewares Association (IHA) in Chicago and organizes IHA juries for annual student design competitions. She also curates many historical houseware exhibits for a number of museums, design firms, and private organizations. On her own time and at her own cost, she interviews aging industrial designers and surviving family members of those who have passed, in order to document industrial design history.

Another colleague, Hampton Wayt, an accomplished design historian and fine arts dealer, graciously provided me with rare images of a 1930 Westinghouse prototype rotating refrigerator concept, and of late 1930 Westinghouse products. Hampton is a dedicated researcher who has acquired the personal records of Donald Dohner, prominent designer and educator of the 1930s, credited by the Industrial Society of America (IDSA) as the father of industrial design education.

IDSA, the single national organization of the profession, generously granted me permission to use images from its publications and those of former organizations, depicting refrigerator designs of the 1950s and 1960s. Thanks to Karen Beruba, Senior Creative Director, and Daniel Martinage, IDSA Executive Director. In 1965, IDSA was formed from the merger of three industrial design organizations: the American Society of Industrial Designers, the Industrial Designers' Institute, and the Industrial Design Education Association.

Chris Hunter, director of miSci, the Museum of Innovation and Science in Schenectady, New York, formerly the Schenectady Museum, was kind enough to give me permission to use a number of images from the museum's extensive General Electric archives. General Electric was instrumental in the development of the first successful domestic refrigerator in 1927.

My son-in-law, Robert E. Schott, of Cranford, New Jersey, is an active practitioner of snow sculpture in his front yard when winter snows come, and has provided me with an example of his handiwork to illustrate how ordinary snowmen can evolve into artistic celebrations of natural snow. His 92-year-old father, Joseph Schott, provided his painting of a typical window card still used in the 1940s to order block ice from the local iceman. I am also grateful to Jimmy Scichilone for information regarding the 1953 Philco "V handle" refrigerator.

Preface

Not long ago, my 16-year-old refrigerator had a problem with its automatic ice-maker, and although it was easily repaired, it reminded me that I hadn't thought much about my refrigerator since I bought it. In human years, my trusty refrigerator would be about 80 years old, so I was appropriately impressed with its unfailing service, much more reliable than my washing machine, dryer, or dishwasher, which barely make it to middle age. Only stoves seem to survive refrigerators.

Not that I am an authority in the secret lives of appliances, but I have designed enough of them in my day to know that they are much more than just the metal, plastic, and nuts and bolts that hold them together: they are the physical representation of human ideas, human engineering, human manufacturing, and human design. As such, I have learned to regard their individual histories as stories of human vision, experimentation, and triumph over nearly insurmountable odds. This book is as much about the humans who contributed to the technology as about the technology itself.

Introduction

Many of us take our electric refrigerators and freezers for granted because few of us can recall being without one. But before 1922, only 5,000 homes in the entire United States had an electric refrigerator. Most other homes still had what were then called iceboxes, which had to be replenished with natural ice blocks delivered to the door by husky icemen, a regular domestic service that had become common since the mid–1850s, over 70 years earlier.

In our modern age, when technology provides us all with scores of conveniences for easier living, instant communication, unlimited travel, and continuous entertainment, it is easy to forget, or never know, the long struggle over centuries, by countless scientists, engineers, and designers, that made each of these products possible. More recent technologies such as iPhones and iPads seem to be developed in a matter of a few years, only because they are simple incremental improvements in a mature personal computer technology that has evolved and been refined over the last 40 years. In the case of refrigerators, it was 210 years ago that the concept of mechanically produced, artificial refrigeration was first imagined. For centuries before that, the human desire to remain comfortably cool in the hottest climates, to cool drinking water, and to preserve perishable foods resulted in numerous discoveries of natural cooling principles and the clever use of natural snow and ice. These discoveries ultimately led to scientific and mechanical principles that duplicated nature in the production of artificial cold.

The history of refrigeration is a fascinating story of human determination to stay cool, and of the many talented men and women who spent their lives working tirelessly to fulfill this desire, sometimes failing, but eventually succeeding far beyond expectations. The story covers a number of general historical eras and topics. First was the pre-industrial era, when humans sought to keep cool though natural means, including the use of natural ice and snow. Second, mostly in the 19th century, came the scientists and engineers who invented huge mechanical machines to make artificial ice and to cool public spaces, and by the late 1920s, reduced their size and cost to make possible small, affordable, domestic refrigerators and air conditioners for use in the home (99.5 percent of U.S. homes have at least one refrigerator and one air conditioner). Third were the designers who styled these products in the new fashions of the 1930s and 1940s to become the modern appliances with which we are familiar today, and which are constantly being improved mechanically, functionally and stylistically. Fourth, and finally, are the many benefits that refrigeration provides today to humankind, the potential danger to our environment it may pose, and future refrigeration technologies. While refrigeration is partially blamed for "global warming," its primary objective is, and has always been, "global cooling."

Among the interesting stories about refrigeration in general, readers will learn how ancient Persians invented a natural home air-conditioning system; how Romans cooled their wine with snow; how ice cream was invented in 1733 and became popular in 1794; how Ben Franklin made artificial ice; how natural ice from New England ponds was transported by ship and sold profitably throughout the world; how mechanical ice-making transformed the South from agriculture to industry after the Civil War; how breweries sustained the fledgling refrigeration industry; how Australian sheep led to successful worldwide shipment of frozen meat; how a French Cistercian Monk brought refrigeration to America's kitchens; how most early electric refrigerators were originally manufactured by automotive companies; how modern kitchen design evolved; how fashion influenced the sale of home appliances, including refrigerators; how Albert Einstein became a refrigerator inventor; how the frozen food industry was founded by a taxidermist; how refrigerators contributed to the fall of the Soviet Union; and how refrigeration saves lives and preserves the dead.

I hope you will enjoy reading this cool story as much as I enjoyed researching and writing it.

Chapter 1

Natural Cooling and Ice

Because this book is largely about being comfortably cool, it would enhance your reading pleasure if at least some of you are sitting in a comfy chair, in an air-conditioned home at a pleasant temperature of 76 degrees and a pleasurable 50 percent humidity, with a glass of iced tea (or other suitable beverage) within reach, cooled with ice cubes (note: they aren't usually "cubes" anymore!) dispensed directly into your glass by your refrigerator simply by pressing your glass against a lever in the door, while the temperature outside is pushing 110 degrees and dripping with 95 percent humidity. If you are not blessed with these amenities, at least you may be able to recall such a pleasant time in your recent past. Most of us today are accustomed to cooling comfort when we need it or want it. We leave our air-conditioned homes and drive in our air-conditioned cars to air-conditioned offices to work in comfort, or to a supermarket, where we buy perishable meat, fish, produce, fruit, and milk that has been preserved or frozen in air-conditioned or freezer trucks, trains, and planes arriving from around the world. We deposit these in home refrigerators or freezers to use at our convenience, days, weeks, or months later. And we finish the day in air-conditioned restaurants or theaters. Lucky us!

Only a few of us older folks can remember summers in the 1940s, only two generations ago, when we tossed awake in bed at night, perspiring because it was so hot and humid, or had an electric fan running night and day to partially combat the oppressive heat but certainly not the humidity. We bought fruit, vegetables, and meat only from local providers, to insure that it was fresh. Only if we lived near the ocean or freshwater lakes was seafood available at all. Each morning, the milkman delivered fresh, cool milk, cream, or butter to our doorstep, and each week or so the iceman brought ice to fill the icebox, which kept such perishables reasonably cool. If we happened to have a refrigerator and wanted an ice cube, we had to pull out the tray and hit it violently on the edge of the sink to dislodge the cubes, frozen like rocks to the tray dividers. Cardboard paddle fans on a wooden stick, with printed advertisements by local merchants, were readily available in churches for hot Sunday services, but they were of limited benefit. The local pool, lake, creek, beach, or, as a last resort, the garden hose, were the only effective ways to keep our bodies cool outside on hot summer days.

But this was nothing new. Such discomfort from extreme heat has been plaguing humans for millions of years, ever since *Homo sapiens* evolved in the tropical savannahs of Africa. There's an anonymous rhyme that expresses it well:

Man is a funny creature.
When it's hot he wants it cold.
When it's cold he wants it hot.
Always wanting what it's not.
Man is a funny creature.[1]

Fortunately, evolution provided humans with a marvelously effective natural cooling system. Unlike most other mammals, we have a minimum of body hair. This allows us to be cooler than our mammalian cousins with heavy fur coats. Anthropologists still debate why this should be so, but some are convinced that about two million years ago, our early ancestors lost their fur and gained special glands in the skin that promoted sweating, and at about the same time, they were able to run for long distances, actually able to run on two legs longer, if not faster, than the four-legged animals they diligently pursued as protein food sources. Our loss of body hair allowed perspiration from thousands of pores in our skin to cool our body by natural evaporation. If we are physically active, we experience this cooling phenomenon frequently.

Technically, such natural evaporation is the changing state of a liquid (water) into a gas (water vapor), a process that inevitably results in a cooling effect, and which is a basic scientific principle fundamental to all our modern mechanical devices that keep us, and perishable food, cool. Our naked ancestors were not aware of scientific principles, but they did notice that moving air, for example a breeze, increased the cooling effect of evaporation on the body. One of the early tools they probably devised was a banana leaf or other vegetation, to fan them when overheated.

As early technology became more sophisticated, banana leaves or other vegetation were replaced with more attractive, durable, and transportable handmade fans. By the 4th century BC, fashionable Greek ladies were using elaborately decorated hand-held, rigid fans called *rhipis,* which vaguely resembled palm fronds. Later, in 2nd century BC China, woven bamboo or palm leaf fans, consisting of a Ping-Pong-paddle-like disk attached to the side of a vertical, hand-held stick, resembling a round flag, was called a *pien-mien*, meaning "to agitate the air," and that's exactly what it did. Chinese inventor Ding Huan, during the same era, decided that bigger was better and developed a rotary fan with seven huge wheels, three meters in diameter, that were manually powered, presumably by slaves.

In backward Christian Europe, fans were unknown until about the 6th century AD, when ceremonial fans called *flabella* were used during religious services to drive insects away from consecrated bread and wine. They are still used for the same purpose by Eastern Orthodox and Ethiopian churches. The classic folding fan was invented in Japan in about the 8th century AD, when only ping-pong-paddle-shaped, fixed, hand fans had been previously known. The folding fans, which pivoted open to a semicircle, were of two types: one was made of cypress wood blades bound by a thread, and the other had a frame with fewer blades covered with Japanese paper, which conveniently folded with zigzag patterns into compact, unobtrusive sticks.

By the 13th and 14th centuries, hand fans from the Middle East were brought back to Europe by Crusaders, and in the 15th century, Portuguese traders brought folding fans to Europe from China and Japan. These fans used bone or ivory sticks that had leather leaves with lace-like cut-out designs, which were slotted onto the sticks rather than glued on like later folding fans. Fan use extended to the highest nobility. Queen Elizabeth I (1533–1603) of England can be seen in formal portrait paintings carrying both folding and rigid fans, often decorated with feathers and jewels. Folding fans became quite popular in Europe in the 17th century, and especially in Spain in the 1800s, where the fan became part of the feminine culture; women used it in various positions over parts of the body, open, closed or waving, as a secret, unspoken, code for courting purposes, a sort of sexual semaphore, called *abanico* in Spanish.

Folding fans became high status symbols on a par with elaborate fashion accessories. They displayed painted leaves, often with a religious or classical subject, or with elaborate floral designs. In the middle of the 18th century, fans were being made throughout Europe by specialized craftsmen, and were decorated and painted by artists. They were imported to the American colonies, along with those brought from China and Japan by the East India Company. Inventors started designing mechanical wind-up fans, similar to wind-up clocks, to fan automatically.

Aside from their secondary functions in fashion, flirting, or decoration, fans were, after all, merely practical devices for creating the cooling effect of air movement on the skin, which increased the evaporation of even mild perspiration. Most fanning by humans necessarily occurred inside closed houses or other structures, where there was little or no natural air movement or breeze, but early tropical cultures soon learned that if they constructed their houses cleverly, they could bring the prevailing natural breezes inside to ventilate and cool the interior more or less automatically.

In Egypt, as early as 1300 BC, residences were designed with wind-catchers or roof ventilators called *malgaf* or *mulguf*. In Dubai wind towers were called *barjeel* or *badjeer* and rose five to fifteen meters above the buildings, open on all four sides. They directed prevailing breezes from any direction downward to cool spaces below. In early Persia (now Iran), people oriented large window openings, *iwanis*, in the direction of beneficial air movements.[2] Later, they built wind-catchers called *Bâd gir*, which created natural ventilation and cooling in the interior. Essentially, a wind-catcher is a tower on the roof of a house that is divided into quadrants, each facing a different cardinal direction, so that no matter what direction the wind comes from, the tower directs the wind downward into the living quarters, thus providing natural ventilation, a basic principle of air conditioning. Many of these structures are still being incorporated into buildings in Egypt and the Middle East today and serve exactly the same purpose.

But in these regions, where the air is particularly dry, such ventilation alone was not, and still is not, enough. To provide significant cooling, it had to be combined with water evaporation, whether from a natural or man-made source. Humans long ago discovered that the evaporation of water on the body, for example after rain or when emerging from a dip into the local river, accelerated significantly the effect of cooling. Much of the primitive technology of cooling developed in two of the hottest tropical countries, Egypt and India, where water was usually available and useful as a cooling medium. Some people cleverly applied this principle of water evaporation to cooling their homes, as ancient Egyptians did, by hanging reeds moistened with trickling water in their windows, thus cooling the natural breezes blowing through. Ancient Romans used their spectacular aqueducts to circulate water through the walls of certain houses to cool them, just as they used fires under other buildings to warm them in winter. In the 8th century AD, Chinese Emperor Xuanzong of the Tang Dynasty had a "Cool Hall" built in his imperial palace, which had water-powered fan wheels for air conditioning as well as rising jet streams of cool water from fountains.

Persians evolved the technique of adding water to the wind-catcher systems into a highly sophisticated art form. In typical applications, located in the basement or lower level of a home, a shaft was dug down vertically into a subterranean, horizontal water passage called in Iran a *qanat* (pronounced "canat"). The Romans, who translated the word as *canali*, adopted the term, and English-speaking people modified the Roman term to *canal*. The *qanat*, usually dug by manual labor, tapped into a water table, often near the foot of mountains, and delivered the water by gravity through a gently sloping tunnel to exit the surface downhill. Connected to the *qanat* were a series of man-made, vertical, well-like shafts, dug vertically from a house basement or lower floor into the *qanat*, and another nearby vertical shaft dug outside the house to allow warm air into the *qanat*, which was cooled from the water evaporation in the *qanat* and forced by low pressure into the house, and after cooling the house, exhausted through the leeward side of the wind-catcher on the roof. Such a system is referred to as an evaporative cooler system, and had the additional benefit of adding moisture to the dry air, a significant comfort to building occupants in dry climates. Today, Iranians have replaced the wind-catcher with an evaporative cooler (*Coolere Abi*), and mechanical evaporative coolers based on this same principle, often used in the American West, are called swamp coolers, desert coolers, or wet-air coolers.

Just as evaporation of water as a primitive cooling mechanism has been used for centuries, it is still used today in many applications such as those canvas desert water bags one often sees hanging on the front of cars driving in the Southwest. They hold one or two quarts of water, which seeps through the porous canvas and evaporates in the rushing air to keep the water refreshingly cool throughout the day, even in 100-degree-plus temperatures. More to the subject of this book, evaporation, the process of any liquid naturally changing into a gas, is the keystone of mechanical refrigeration with which we are all familiar. Our complex refrigeration technology of today is simply a refinement of this simple, fundamental, scientific principle.

Simple evaporation of water could not only produce effective cooling, but could actually make artificial ice. Egyptian tomb paintings from 2500 BC show slaves vigorously fanning large storage jars. The jars were made of porous clay, so that a small amount of liquid would seep through and evaporate on the surface with fanning (just like today's desert water bags mentioned above).[3] Protagoras (490 BC–424 BC), the Greek philosopher, described a 5th century BC Egyptian method of cooling water to make ice:

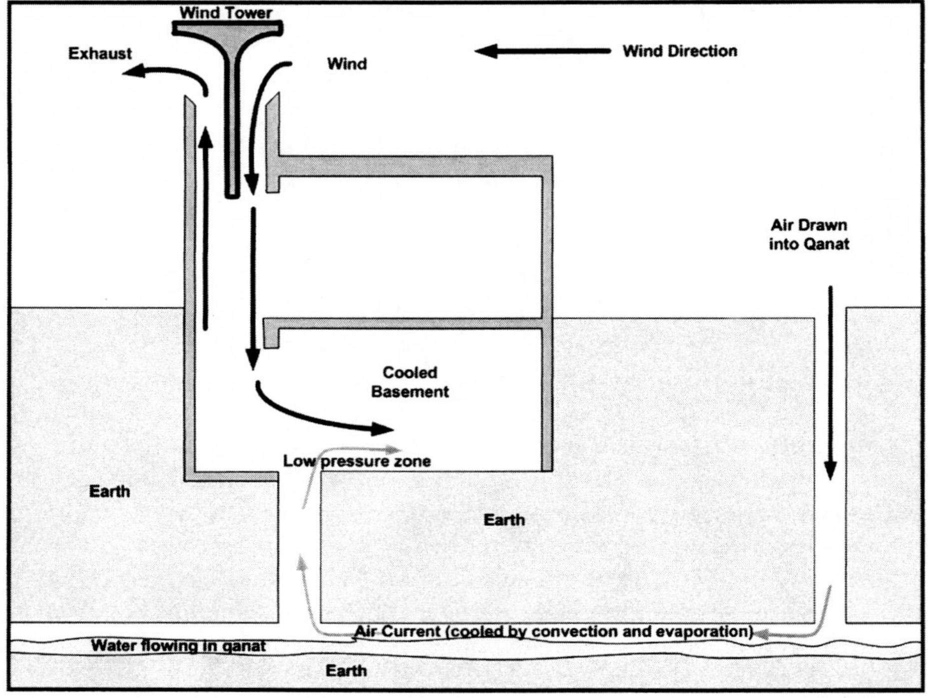

Persian wind-catcher and qanat evaporative cooling system.

> For during the day they expose it to the sun, and then at night they skim off the thickest part which rises to the surface, and expose the rest to air, in large earthen ewers [jugs], on the highest parts of the house, and two slaves are kept sprinkling the vessels with water the whole night. And at daybreak they bring them down, and again skim off the sediment, making the water very thin and exceedingly wholesome, and then they immerse the ewers in straw, and after that they use the water which has become so cold as not to require snow to cool it.[4]

One has to wonder what exactly the "sediment" was. It may have been icy slush, such as produced in a similar method used in ancient southern India in the 4th century AD In this method, shallow, porous clay pots filled with boiled, cool water were lined up in rows in shallow trenches, covered with straw and exposed to drafts on winter nights. Under favorable conditions, thin ice, called "hoogly ice," would form on the surface for harvesting and sale. Such ice was only available in limited quantities and was considered of poor quality, because it resembled slush. The same process was being used in Allahabad, India, as late as 1843. Even today, the simple

Evaporative ice manufacturing system, Allahabad, India, 1843.

technique of cooling drinking water in unglazed pottery jugs, by evaporation of seepage on the surface, is probably the most widely used process of cooling water in undeveloped tropical countries of the world, where mechanical refrigeration is unavailable.

Primitive tropical cultures, despite all their ingenuity in devising artificial means of cooling, knew that nature's perfect coolant, ice and snow, existed on the tops of mountains during the winter, or sometimes all year, just for the taking. There is a long history of humans collecting this natural resource, despite the difficulty of doing so, from the Andes, the Himalayas, the Alps, and many other mountain peaks located in warm climates. The primary advantage of such natural materials, of course, was the obvious effective preservation of perishable foods, the cooling of liquids, or the cooling of humans. The primary challenge was in figuring out a way to preserve this ice or snow before it melted.

As early as 1700 BC, Zimri-Lin, ruler of Mari (a kingdom in northwest Iraq) boasted of building a *bit shuripim* (icehouse): "which never before has any king built on the bank of the Euphrates." The ice was imported from nearby mountains and probably stored in a pit.[5] An ancient poem, the *Shi Ching,* from about 1175 BC, describes harvesting ice for a pagan ceremony honoring the annual return of spring:

> In the bitter cold days of the second month, the ice flows
> are as hard as rocks;
> The axes ring with a merry clang, as we hew out the ice in blocks.

The third month comes,
Ere the thaw begins,
The ice in caves we store;
Then our ploughs made ready to till the land, for Spring
is at hand once more.
When the hot days come, we must open the cave
Wherein we have stored our ice;
But first to the gods, at the dawn of the day,
A lamb we must sacrifice.[6]

The process of saving winter ice and snow for summer use was pretty simple, but ingenious. According to the Greek classical food writer Athenaeus, Alexander the Great (356–323 BC), during his campaign in northern India in about 330 BC, "dug thirty cooling pits which he filled with snow and covered with oak boughs. In this way, snow will last a long time."[7] This is another basic scientific principle at the heart of any refrigeration process—insulation. Ice and snow, like water, evaporates naturally, and as it melts and evaporates, it cools the immediate air around it, which is its primary advantage as a coolant. But if it is protected by insulation, such as the oak boughs above, or by other materials, from the warmer ambient air, the natural process of melting and evaporation is slowed considerably.

In the Persian Empire in 400 BC, an above ground, several-story-tall icehouse of adobe, shaped like a pointed dome, was called a *yakhchal*. The first emperor of China, Shih Huang Ti, at Hsienyang, in about 220 BC, built an ice pit that was dug 43 feet below ground level and lined with large terra cotta rings about 3 feet high and 5.5 feet in diameter.[8] The temperature beneath the upper 6 meters (20 feet) of earth maintained a nearly constant temperature of 10°C to 16°C (50°F to 60°F).

One of the early uses of ice and snow was to cool wine. As any wine aficionado knows, wine is best chilled in both storage and drinking. The Romans, who never invented any method of creating ice from scratch, nevertheless loved their chilled drinks and even chilled baths called *frigidaria*, both using snow. No Roman banquet would have been complete without lavish quantities of ice or snow to sprinkle into their wine, even though some of the elders regarded it as a sign of decadence in the day. Seneca, the tutor to the young emperor Nero in the 1st century AD, commented sarcastically: "No running water seemed cold enough for us." He described the common practice of using cold water for reviving Romans who had drunk themselves in a stupor. He raged against "the shops and storehouses for snow, and so many horses appointed to carry the snow," as a gross waste of time and effort.[9] Seneca (apparently somewhat of a grouch) also complained of youthful Roman decadence: "You see only skinny youths wrapped in cloaks and mufflers, pale and sickly, not only sipping the snow, but actually eating it and tossing bits into their glasses lest they become warm merely through the time taken in drinking."[10]

Roman doctors tended to agree with Seneca, however, believing the practice of mixing dirty snow into drinks was the cause of serious internal disorders. This did not discourage many Romans, certainly not extravagant emperors. Emperor Nero (37–68 AD) himself combined the snow with fruit toppings, but also developed a more sanitary method for ice water by immersing glass containers filled with boiled water directly into the snow. Pliny noted that Nero's method ensured "all the enjoyment of a cold beverage without any of the inconveniences resulting from the use of snow." The Roman Emperor Elagabalus (203–222) told of Emperor Varius Avitus in about 220 AD, who ordered mountain snow to be formed in mounds in his garden so natural breezes might be cooled—an early attempt at domestic air conditioning.[11] Snow tanks were discovered in the cellars of Hadrian's villa in Tivoli, built at about the same time.[12]

Where did all this snow come from in sunny Italy? Donkey trains carried it into the cities, often over considerable distances, from Mount Albanus (now Monte Cavo, an old volcano and the second highest mountain of the Alban Hills, near Rome). After it arrived in the cities, it was either sold directly in "snow shops" or stored in deep underground pits, covered with straw or tree boughs, until needed or sold. It was used to directly cool beverages, or used for cold baths. The top layer of snow in storage might melt, but it would freeze again as it seeped down, and eventually the weight of the snow column would turn the lowest level into solid ice. This is how the snow salesmen manufactured ice, which they could sell at a much higher price than snow.

Old Seneca complained about this, as well: "Even water has a varying price." In old Rome, ice and snow often cost more than wine. Pliny the Younger complained to his friend Septimus Clarus for failing to show up for a lavish dinner arranged for him, specifically about the waste of "sweet wine and snow; the snow I'll most certainly bill you for and at a high rate—it was ruined in the serving."[13] The Greeks matched the Romans in their use of snow to cool beverages. The Greek Strattis said: "No man would prefer to drink hot wine; rather one likes it chilled in the well or mixed with snow."[14] Archeologists found a Greek wine cooler from the 6th century BC in Vulci, central Italy. On the outside, it looked like an ordinary large jar, but on the inside was a central container for the wine, surrounded by another chamber that could be filled with ice or snow.

But wine was not the only thing chilled by snow. There are records of an 8th-century Baghdad caliph using snow packed between the walls of his summer villa for cooling. In 400 BC Persia, in addition to the practical preservation of food, snow was used as summer treats for centuries by pouring grape juice over it, creating a sort of sorbet. Snow was mixed with saffron, fruits, and various other flavors. Since that time, *faloodeh* (originally *paloodeh,* meaning smoothly filtered), a Persian cold dessert, has been prepared, consisting of thin noodles made from corn starch mixed in a semi-frozen syrup made from sugar and rose water, often served with lime juice. It was served to royalty during summers and is still a traditional dessert in Iran and Pakistan. Other ancient cultures also used ice for milk-based desserts. In China in about 220 BC a frozen mixture of milk and rice was reported to be a summer delicacy. Arabs were known to use milk as a major ingredient in cold desserts. In the 10th century such desserts were popular in Baghdad, Damascus and Cairo. Produced from milk or cream, often with yogurt, the dessert was flavored with rose water, dried fruit, and nuts. The recipe was probably based on earlier Arabian recipes.[15]

Legend says that the Italian merchant Marco Polo (1254–1324) returned from his travels in China in the 13th century with a recipe for ice cream obtained from Kublai Khan (1215–1294), the emperor of China, who enjoyed it and kept it as a royal secret for years before revealing it to Marco. The preparation secret is believed to have been the evaporation of brine (salt water) and ice, which would have lowered the freezing temperature below normal (32 degrees Fahrenheit). As a matter of fact, when Dutch-German physicist Daniel Fahrenheit (1686–1736) in 1724 invented the temperature scale named after him, he discovered that using a brine solution with ice lowered the temperature to 0°F before freezing occurred. We are all familiar with this principle of salt lowering the freezing temperature of water. This is why fresh water freezes before salt water, and why we salt our roads in winter, or sprinkle rock salt on our icy front steps—to melt ice and snow that froze at 32°F, but with salt, will not freeze again until or unless the temperature drops to 0°F.

Whether the Marco Polo legend is true or not, in the 16th century it was the Italian philosopher and medical student Marcus Antonius Zimara (1460–1532) who in his *Problema 102* in 1525 recommended the mixture of snow and saltpeter (ammonium nitrate) as an effective cooling medium.[16] Another Spanish physician practicing in Rome in 1550, Blasius Villafranca, discovered that by adding potassium nitrate to ice or snow, a lower freezing temperature could be reached; and, as a matter of fact, he may have been the first to use the term "refrigerate" in his description of the process. In 1559 Levinus Lemnius of Antwerp noted that with this method, wine could become so cold that it could be uncomfortable to the teeth. It was observed in 1646 by the Jesuit priest Cabeus that ice could be made by this process, and several Italian authors, Della Porta in 1589 and Latinus Tancredus, a professor at Naples, in his *De Fame et Siti* in 1607, mentioned the action of water and ammonium nitrate with the addition of snow, so as to make ice.[17]

In France and Italy, ice was used to make another summertime beverage with which we are quite familiar. In about 1630, French descendants of Italian liqueur merchants invented an iced drink they named "lemonaid." Later, in 1660, a Florentine Italian, Procope Couteaux, began selling solidly frozen lemonaid. By the end of the 1600s, vendors (called lemonadiers) selling iced liqueurs or frozen drinks were common in France. This trade used large quantities of ice and at the same time, the sale of snow was also a profitable trade in France.[18]

As the age of scientific exploration began in the 17th century, the English philosopher, chemist, physicist, and inventor Robert Boyle (1627–1691) was famous for Boyle's Law (1662), which stated that the pressure of a gas decreases as the volume increases. Boyle also experimented with freezing mixtures of various salts in the 1660s and concluded that ammonium chloride was better than either ammonium nitrate or potassium nitrate for freezing mixtures.[19]

People continued to experiment with chemical freezing mixtures, as described by Johann Beckmann (1739–1811), a German scientific author who coined the word "technology" to mean "the science of trades." He quoted from an event described in the *Argensis*, a 1659 book in Latin by Jean Barclay:

> Arsidas finds in the middle of summer, at the table of Juba, fresh apples, one-half of which was encrusted with transparent ice. A bason [basin] made also of ice and filled with wine, was handed to him; and he was informed that to prepare all these things in summer was a new art. Snow was preserved the whole year through in pits lined with straw. Two cups made of copper were placed one within the other, so as to leave a small space between them, which was filled with water; the cups were then put into a pail, amidst a mixture of snow and unpurified salt coarsely pounded, and the water, in three hours, was converted into a cup of solid ice, as well-formed as if it had come from a pewterer [a craftsman who made pewter utensils]. In like manner, apples pulled from a tree were covered with a coat of ice.[20]

This sounds a bit like a parlor trick, and in the 1600s, it was sure to amaze and delight. But ice and snow were not very helpful to solve a serious problem of life and death for those affected: the freshening of the dangerous air in mining operations. Ventilation, another component to effective conditioning of air, was what was needed. Earlier, we described ancient methods of ventilating buildings with wind-catchers to direct natural breezes inside for comfort cooling, but there were no natural breezes in subterranean mines, ship holds, or large buildings. As mechanical means were developed, such ventilation systems were devised more for health and safety reasons. The earliest such systems were applied only to the most extreme conditions such as mining, where poor air quality and toxic elements crippled or killed miners. Georgius Agricola's *de re Metallica* of 1556 (translated from Latin to English in 1912 by former president Herbert Hoover, then a civil engineer, and his wife, Lou) describes ventilating mine shafts with rotating wooden fans and bellows, powered by men or horses. Later, mechanical ventilation was applied to the more general needs of industrial, commercial, and domestic buildings. One of the earliest of these was a ventilating system designed for the Houses of Parliament in 1660 by Sir Christopher Wren (1632–1723), architect and physicist. Though the system left much to be desired, it was the first large central ventilating system for buildings and has been regarded as the beginning of the history of ventilation,[21] another basic principle of refrigeration.

A major cause of poor air quality in domestic interiors was the extensive use of open-hearth fires and oil lamps. The development of the fireplace flue and operable sash windows, the Western world's oversimplified version of the Middle East wind-catchers, alleviated these problems somewhat. Although sash windows first appeared in 1300, they did not become common until about 1700. The publication of *La Mechanique du Feu* by Frenchman Nicolas Gauger in 1713, and its English translation, *The Mechanism of Fire Made in Chimneys,* by Dr. John Theophilus Desaguliers, was probably the first to include discussions of the ventilating of buildings. Desaguliers showed a fanning wheel at the Royal Society of London in 1734 that he had designed in 1727 for the Earl of Westmoreland to clean foul air out of mines. Desaguliers then used his "fan engine" to ventilate the House of Commons from 1734 to 1736. Using this "fan engine" with its 7-foot diameter and 1-foot-wide blades, one man cranking manually could force 10 cubic feet of air per minute.

In 1758, the Reverend Stephan Hales proposed ventilation in ships by using bellows in his *Treatise on Ventilators*. He optimistically estimated that with a 10-foot bellows worked by two men at the lever, the machine would expel a ton of air per stroke, or six tons per minute. Hales also

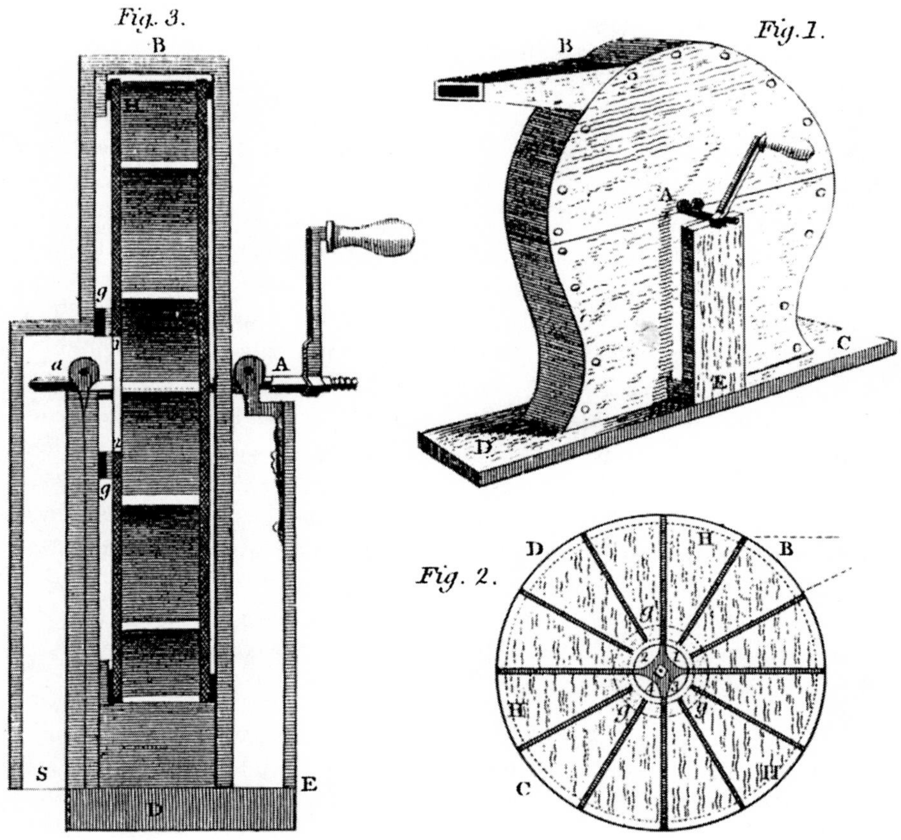

Desaguliers "fan engine," from **Philosophical Transactions** *(abridged), vol. 8, 1735.*

worked on designs for bellows ventilation of the county hospital and county jail in Winchester, the Savoy Prison, and Newgate Prison.[22] As you can imagine, all these devices, although mechanical in nature, were highly labor intensive. As late as 1845, it took two men all evening to ventilate Queen Victoria's (1819–1901) opera box. Of course, all these ventilating systems were without much cooling effect. It was English physicist and mathematician Sir Isaac Newton (1642–1727) who, in his 1701 *Principles of Natural Philosophy*, first described the scientific laws of cooling, stating the relationship between thermal transfer, surface area, and temperature difference. Scientific principles such as these provided a knowledge base for the development of the industrial revolution and for using these principles practically in the physical world.

Most ordinary people were not very interested in scientific principles but were quick to recognize practical results in the physical world. Sometimes ideas were not particularly practical but brought culinary pleasure. Such is the case with ice cream. The ancient process of making ice by chemical means, that is, by the addition of salts to ice, as earlier reported to have been revealed to Marco Polo by Kublai Khan in the 13th century, and demonstrated scientifically by Robert Boyle in the 1660s, is the same scientific principle that made ice cream feasible. Most legends of the origins of early "ice cream" were not about what we know as ice cream, but were actually flavored "iced water" or "water ice." Nevertheless, by 1733, recipes for genuine ice cream were published in England, which included the preparation secret of salt and ice. In 1744, a group of VIPs dined at the home of Thomas Bladen (1698–1780), Maryland colonial governor from 1742 to 1774. Present was a Scottish colonist who provided the first written account of ice cream consumption in America; the dinner was followed by "a Dessert no less Curious; among the Rarities of which it was Compos'd, was some fine Ice Cream which, with the Strawberries and Milk, eat most Deliciously."[23]

It was probably prepared by the "pot freezer method" as it was known at that time, presumably similar to that described by a recipe for ice cream published in London in 1733 in *Mrs. Mary Eales's Receipts*:

> Take Tin Ice-Pots, fill them with any Sort of Cream you like, either plain or sweeten'd, or Fruit in it; shut your Pots very close; to six Pots you must allow eighteen or twenty Pound of Ice, breaking the Ice very small; there will be some great Pieces, which lay at the Bottom and Top: You must have a Pail, and lay some Straw at the Bottom; then lay in your Ice, and put in amongst it a Pound of Bay-Salt; set in your Pots of Cream, and lay Ice and Salt between every Pot, that they may not touch; but the Ice must lie round them on every Side; lay a good deal of Ice on the Top, cover the Pail with Straw, set it in a Cellar where no Sun or Light comes, it will be froze in four Hours, but it may stand longer; then take it out just as you use it; hold it in your Hand and it will slip out. When you wou'd freeze any Sort of Fruit, either Cherries, Rasberries, Currants, or Strawberries, fill your Tin-Pots with the Fruit, but as hollow as you can; put to them Lemmonade, made with Spring-Water and Lemmon-Juice sweeten'd; put enough in the Pots to make the Fruit hang together, and put them in Ice as you do Cream.[24]

Mrs. Mary Eales was confectioner to Queen Anne (1665–1714) of Great Britain. The origin of this "pot method" of making ice cream is lost to history, but probably goes back many years. Ben Franklin, George Washington, and Thomas Jefferson were all known to have regularly eaten and served ice cream. Jefferson and Franklin spent considerable time in Europe and, as connoisseurs of fine European cuisine, would have been not only aware of this delicacy, but inquisitive enough to discover how to make it. Jefferson was one of the first to serve ice cream at a state banquet in the White House. According to *Godey's Lady's Book* in August 1850, by 1794, ice cream was so popular in Europe that the famous composer Ludwig van Beethoven (1770–1827) complained: "It is very warm here. The Viennese are afraid that it will soon be impossible to have any ice cream, for as winter is mild, ice is rare."[25] In the same 1850 magazine, the editor commented on hand-cranked ice cream freezers, which had been invented in 1843: "Ice cream had become one of the necessities of life. A party without it would be like a breakfast without bread or dinner without a roast."[26]

First Lady Dolley Madison (1768–1849), who would graciously serve ice cream at her husband's inaugural ball in 1813, is often associated with the early history of ice cream in the United States; in fact, she had an ice cream brand named after her in the 1950s (Dolley Madison brand). Dolley had heard about a Mrs. Jeremiah Shadd (known as Aunt Sally Shadd), a free black slave, who had achieved legendary status among Wilmington's (Delaware) free black population as the inventor of ice cream. She had opened a catering business with family members and created a new dessert sensation made from frozen cream, sugar, and fruit. Dolley went to Wilmington to try it and enjoyed it so much that it became part of the menu at the inaugural ball in 1813. Augustus Jackson (b. 1808), an African American who served in the White House as a chef during the 1820s, left his job at the White House to return to his hometown, Philadelphia, and establish a successful ice cream business. Around 1832, he not only created multiple ice cream recipes, but also invented a superior technique to manufacture ice cream. He distributed it in tin cans to Philadelphia's many ice cream parlors.[27]

Around the New York area in the late 1700s, the hot summers and growing economy began to increase public demand for ice, creating a small-scale market for farmers who sold winter ice from their ponds and streams to local city institutions, perishable food vendors, and families. In New England, ice was an expensive product, consumed only by the wealthy who could afford their own ice houses; but around 1800, there were quite a few of these, whose owners cut ice and filled these houses with it during the winter months. Aside from ice, of course, there were many ways to preserve food that had been used since prehistoric times, such as drying fish, meat, fruit and vegetables. At the time, agriculture was the primary occupation in most of the world, and preservation of food was an essential part of farming. Making wine from grapes, cider from apples, and jams from fruit were all common forms of preservation. Cuts of meat could be cured in salt, or smoked in the many smokehouses in farms around the country. These methods have a number of drawbacks, however. They destroy many of the nutrients in food and their flavors are changed so that they cannot pass for fresh. Refrigeration was the only form of food preservation that did not change the food's taste.

In fact, farmers did have an effective form of refrigeration without ice that had been utilized for centuries. Farmers were always faced with the preservation of highly perishable milk, cream and butter, as well as vegetables and meat. Many farmers accomplished this with springhouses, where cool, fresh spring water at below-ground temperature (about 50 to 60 degrees) was readily available year round. Usually built in stone over a natural mountain spring, such small buildings without windows were often built partly below ground to take advantage of the more constant, cooler below-ground temperature. They were constructed so that the cool water from the spring flowed constantly over a shallow, wide, sloping concrete trough, where the water depth was only a few inches. Into this steady flow the farmer placed his tins of milk, cream, butter, and the like, which were preserved for weeks in this manner. On shelves on the interior walls, fresh produce and meat could be stored and preserved, at least for short periods of time, until sold, canned, smoked, or cured. This ingenious system had been used for millennia and was totally practical. It also served as a source of fresh, cool drinking water. Farmhands could stop by to cool off from the hot summer heat, and refresh themselves with the inevitable tin cup hung beside the spring. If the farmer had a pond from which to harvest winter ice, the springhouse could have served as an icehouse, as well, to make the springhouse even more efficient as a refrigerator. From the 16th century onward, wealthy Europeans and, later, Americans had built ice houses; insulated structures above ground to store ice and snow gathered during the winter, and used during the summer to cool drinks or preserve perishable food. Such icehouses would be built and used for commercial purposes well into the 20th century.

Household appliances of the mid–1800s included an "artificial freezer" to make ice cream, invented by Nancy M. Johnson of Philadelphia, who patented it (U.S. Patent 3,254) in 1843, then sold her patent to William Young, who marketed it as the Johnson Patent Ice-Cream Maker. It contained an outer bucket, an inner metal cylinder, a dasher (paddles), and crank; a basic design still widely used today. The metal cylinder was filled with cream and sugar flavored with vanilla or fruit, placed in the tub, and

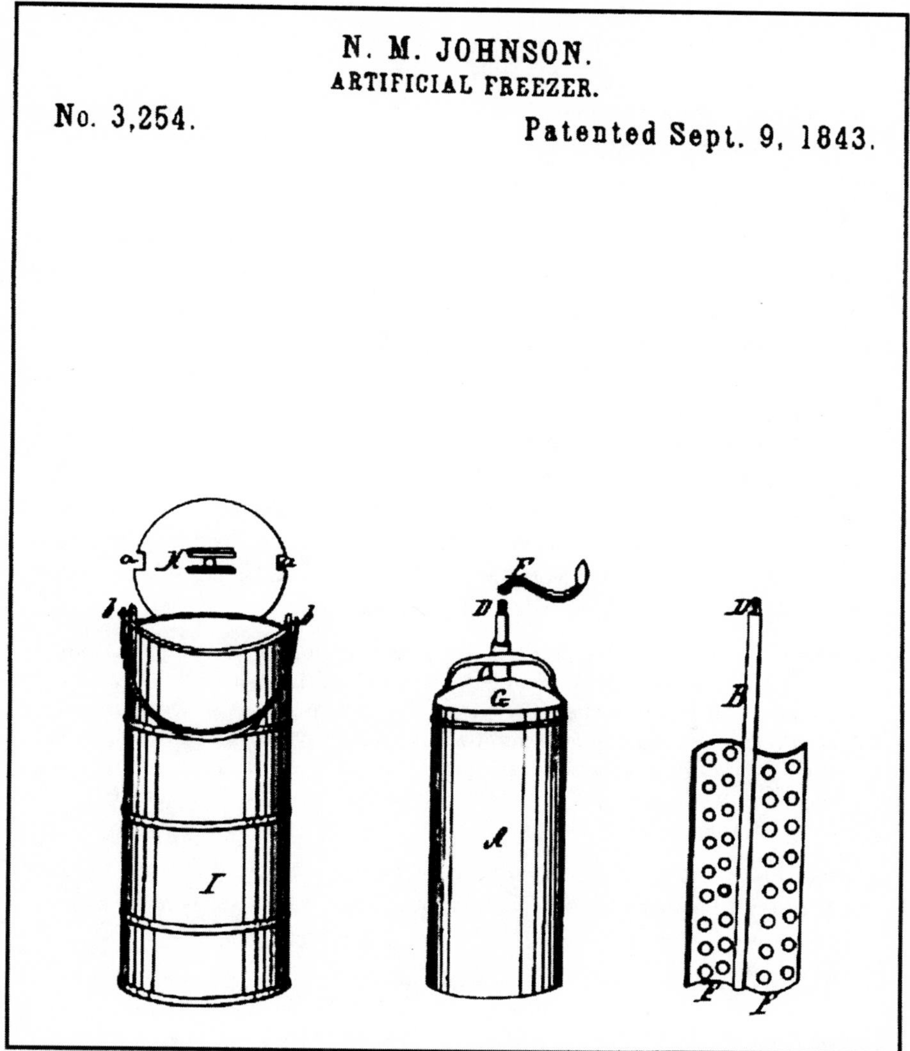

Nancy Johnson's "artificial freezer" to make ice cream. U.S. Patent No. 3,254.

surrounded by ice and rock salt. Inside the cylinder was a rotating dasher, metal paddles attached to a spindle, which extended through the lid and was driven by a gear on top operated by a hand crank. The crank turned the paddles inside the container, and when turned vigorously for some time, the cream mixture became ice cream. It was a task demanding considerable manual labor and time, but all who shared the results agreed that it was well worth the effort. The next patent for an "ice cream freezer" was U.S. Patent 6,865 in 1849 by Goldsmith Coffeen, Jr., which added a hand-operated bellows to blow air over the ice into the cream mixture instead of rotating the cylinder with a crank. Nothing further is known of this idea, but it is doubtful that it was less physically exhausting than a crank, and probably took much longer than the half hour or so with a crank.

To make ice cream, you needed ice, and there was none in the South, where there was no pond ice to harvest, as there was in the North. Some ships occasionally transported ice from New York and Philadelphia to wealthy Southern plantation cities such as Charleston in South Carolina, using it as ballast on the trip. Ice was obviously more profitable than rocks, the normal ship ballast. What was needed for these early users of ice shipped from the North was some sort of an insulated container to delay the melting of the ice, and also to hold and cool perishable goods, the primary purpose of ice. The term *refrigerator* was coined by a Maryland farmer and civil engineer, Thomas Moore, in 1800, for his design for transporting butter from rural Maryland to Washington, D.C. He patented the device in 1803, and the patent stated its basic concept:

> The principle of this refrigerator is the application of ice in such a manner that it can receive little or no heat, except from the matter to be rendered cold or deprived of its heat. The trials hitherto made have been made by an internal vessel of tin, surrounded by an external vessel of wood, so as to leave a small interstice between them filled with ice. The external vessel [has] a coat of woolen cloth lined with rabbit skins, the fur included between the cloth and the pelt.[28]

The original letters of patent, signed by Thomas Jefferson, then president, and James Madison, then secretary of state, were destroyed in a fire in 1945. Moore's nephew described the actual construction:

> The first was made of a size to be carried on horseback. The later ones were large and made in a different manner. Two square cedar boxes, one of a smaller size to be placed in the larger, and this space filled with pulverized charcoal well packed in, a tin box fastened to the inner side of the lid contained the ice, and the whole covered with coarse woolen cloth. Since this was made to be used by people who had icehouses, and there were not that many, other than Thomas Jefferson, some of his cabinet members, and people of wealth, such refrigerators were not extensively used. The patent expired in 14 years. It was not renewed, so the public received the benefit of it.[29]

The next evidence of a refrigerator is U.S. Patent 750, granted May 25, 1838, to Henry V. Hill, for an "improvement in constructing refrigerators."

Today, we would call these "iceboxes," not refrigerators. We reserve the term "refrigerator" for the electro-mechanical devices in our kitchens with which we are most familiar, but they did not become common until the 1930s, and for a generation after that, many older people still called them "ice boxes," the term for the earlier device. Technically, the definition of a "refrigerator" is any kind of enclosure (box, cabinet, room, or house) whose interior temperature is kept substantially lower than the surrounding environment. So back in the 19th century, they were quite correct in calling them "refrigerators."

However, the small, local market use of natural ice was about to expand around the world. Frederic Tudor (1783–1864), an ambitious 23-year-old Boston businessman and merchant, founded the ice trade by initiating a global distribution of ice in 1806. Inspired by his brother William's idea that ice could be harvested from the ponds of Massachusetts and exported to the warm climates of the Caribbean, he invested $10,000, confidently established the Tudor Ice Company, and bought his first brig, *Favorite,* to carry ice from his father's farm in Saugus, Massachusetts, to the Caribbean. In February 1806 the brig sailed 1,500 miles from Charlestown, Massachusetts, to Martinique. At the time, the business community considered Tudor, at best, something of an eccentric, and at worse, a fool. The *Boston Globe* found the idea to be somewhat humorous: "No Joke. A vessel has cleared at the Custom House for Martinique with a cargo of ice. We hope this will not prove a slippery situation."

While Tudor was making preparations for the shipment, he sent his brother William and his cousin, James Savage, ahead to obtain a monopoly from the various governments of the islands. This effort failed, but he was given permission to sell the ice directly from the ship. Although a considerable amount of ice melted during the three-week voyage, he did manage to sell what remained for $4,500, at 25 cents per pound ($88,500 and $4.92 in today's dollars). In the subsequent year, however, he had severe financial problems when three shipments to Havana in the *Trident* resulted in severe losses due to no facilities existing there to store or distribute ice, a serious oversight.

His first profits were finally realized in 1810 when his gross sales totaled $7,400, then increased to $9,000 ($177,000 today). But he lost most of that to what he called the "villainous conduct" of his agent. Tudor's personal debts exceeded his income and he spent parts of 1812 and 1813 during the war with England in debtor's prison. However, by 1815 he borrowed money to buy more ice and to pay for building a new icehouse in Havana that held some 150 tons of ice in storage. In 1816 he was shipping

Slaves unloading ice in Havana, Cuba, for the Frederick Tudor Company, 1832.

ice to Cuba with increasing efficiency, and he borrowed $3,000 (at 40 percent interest) to import a shipload of limes, oranges, bananas, and pears from Cuba to New York, preserving it with 15 tons of ice and 3 tons of hay as insulation.[30]

It was a disaster, since almost all of the fruit rotted during the monthlong voyage, leaving him only with more new debt. But Tudor was not a man easily discouraged, and he resolutely pressed on, opening new markets in Charleston, South Carolina; Savannah, Georgia; and New Orleans, Louisiana; where ice could be sold for 8 cents ($1.20) per pound. Ice was not always welcomed where it arrived. In about 1820, New Orleans mayor Augustin McCarthy reportedly said when he met the first sailing ship to deliver ice to the city: "Iced drinks chill the innards and make consumptives [tuberculosis patients] of the people who drink them." He then shocked the crew by ordering every block of ice to be thrown into the harbor. Tudor spent the next few years experimenting with different types of insulation, packing ice aboard ships with wood shavings, sawdust, or rice chaff. During this time, the lumber industry was beginning in New England, and sawdust was plentiful and cheap. Ice blocks were stacked together in such insulation like well-fitted masonry.

Tudor constructed icehouses throughout the tropics with insulation that reduced ice wastage from 66 percent to 8 percent, and created customer demand by the Cubans for cold refreshments. By 1812 he was also selling Boston brand "refrigerators" to households, calling them "little ice houses." They were improvements on the 1803 Thomas Moore patented design: made of wood, lined with iron, insulated with sheep fleece, and designed to hold about three pounds of ice. They were an essential part of the ice system because without them, households could not use ice over a period of time without rapid melting. By 1825 Tudor was doing well with ice sales because of significant improvements in harvesting techniques and distribution. Around 3,000 tons of ice was being shipped from Boston annually, two thirds of it by Tudor.[31]

Ice harvesting procedures were tedious, labor-intensive, and fraught with technical problems as well as weather challenges. Ice on the lakes needed to be at least 18 inches thick, as it had to support the weight of men, horses, and equipment. Typically, a crew of 100 men and 30 to 40 horses were required. In New England, holes were drilled in the surface to promote thickening. In addition to the necessary frigid working conditions, ice cutting was typically carried out at night, when ice was the thickest. First the snow had to be cleaned off with horse-drawn scrapers, and the ice tested for suitability. The blocks were initially cut out with handsaws into irregular shapes, but these proved to be difficult to secure on ships, so later, lines were marked off in a grid on the ice over 2 to 3 acres with cutters, to about 2 to 3 inches deep, to define the size of future uniform blocks. The size varied with destination, the largest being for the furthest locations, and the smallest, about 22 inches square, for closer destinations such as the American East Coast.[32]

The blocks were broken off with ice spades from a strip and floated to the shore, where they were trimmed to size as necessary. Men with ice hooks drew the ice blocks up ramps onto platforms, and when full, the platforms were slid onto sledges and into icehouses on the shore. Icehouses were built of pine walls and filled with sawdust to a thickness of two feet. Next was the arduous transportation of the heavy ice blocks, packed in sawdust, from the ponds or icehouses onto special wagons or trains to the ports where ships waited.

Specialized protective equipment, such as cork shoes for the men and spiked horseshoes for the horses, were required, as well as pickaxes, axes, ice saws, grapples, tongs, bars, ice spades, and chisels to handle and work the ice into uniform blocks. Harvesting rights had to be purchased from pond owners, often multiple shore landowners, so the rights had to be divided in proportion to the amount of shoreline owned by various

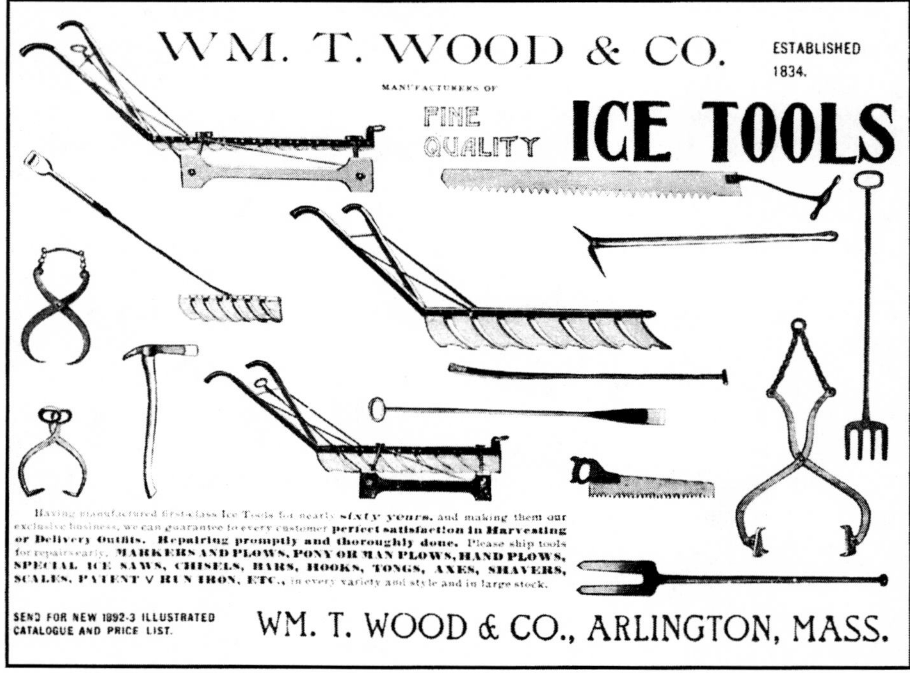

Ice harvesting tools advertisement from* Ice and Refrigeration *magazine, March 1893.

individuals. As a result, property values on Fresh Pond in Cambridge increased from $130 ($3,160 today) in the 1820s to $2,000 ($61,500) per acre by the 1850s. Further, a warm winter could cause seasonal disaster by resulting in no ice at all, or thin ice that could not produce the large blocks required. Such shortages of ice were called "ice famines" in the trade.[33]

Despite these problems, manual labor was the least of Tudor's worries. During ice harvesting time, farmers were idle, and they were happy to earn extra money working in the ice trade. Others were too, such as Boston inventor and businessman Nathaniel Jarvis Wyeth (1802–1856), who started his career in the 1820s as a foreman for Tudor's ice harvesting operations on Fresh Pond in Cambridge, and invented a number of tools that revolutionized the business by increasing productivity and enabling a larger scale, more commercial, harvesting process.[34]

Among these was a horse-drawn ice cutter, resembling a plow, with two parallel steel cutters to mark the ice quickly and accurately, patented in 1829, and a horse-drawn plow with steel teeth to assist in the cutting process. By the 1850s, ice tool manufacturers would be producing catalogs and selling products along the East Coast.[35] Wyeth also invented the above-ground ice houses, with double walls for insulation. They evolved later into huge, multi-storied warehouse structures with steam-powered conveyor belts that pulled the blocks of ice up inclined planes to the desired level, where blocks were slid by gravity into storage areas. Men would then pack them neatly with sawdust into compact towers of ice.[36]

Some brand names became well known. "Wenham Ice" from Wenham Lake in Wenham, Massachusetts, became internationally famous for its crystal clarity, highly desirable for decorative table ice. It was harvested by Tudor's company, the Wenham Lake Ice Company, which opened a storefront in the Strand, London. Every day, workers would put a large block of ice in the window, with a newspaper on the other side so that passersby could read the print through the ice from outside the store. The ice was awarded a royal warrant from Queen Victoria.[37] In 1828, Londoners paid about £100,000 per year for ice.[38]

In 1833, Tudor partnered with fellow merchants Samuel Austin and William Rogers to sell ice to India, which was 16,000 miles away and 4 months' sailing time from Boston. That May his brig, *Tuscany*, sailed for India with 180 tons of ice. When it arrived, 100 tons still survived, which was sold at a profit of $9,900 ($283,000 today). Over the next 20 years, Calcutta became his most profitable customer, generating $220,000

Ice harvesting operation, New York, 1852.

($6,300,000) in profits. He exported additional ice to Madras and Bombay. By 1856, 146,000 tons would be shipped to India from New England. In 1834, he began sending shipments to Rio de Janeiro, with ships returning with cargos of sugar, fruit, and cotton. In 1838, Tudor was granted U.S. Patent 726, *Improvement in the Method of Packing and Stowing Ice*.[39]

Tudor was harvesting ice from dozens of ponds in the Boston area, including, starting in 1842, Walden Pond in Concord, Massachusetts, which became the inspirational home of writer and philosopher Henry David Thoreau (1817–1862) from 1845 to 1847. One cold morning, one hundred Irish immigrant workers arrived at Walden Pond where Thoreau lived in a cabin he had built for himself. In *Walden*, Thoreau noted that the men came

> with many car-loads of ungainly looking farm implements, sleds, ploughs, drill-barrows, turf knives, spades, saws, rakes, and each man was armed with a double-pointed pike staff.... I did not know whether they had come to sow a crop of winter rye, or some kind of grain recently introduced from Iceland.[40]

Thoreau soon learned that they were ice harvesters who had come to harvest blocks of ice from the pond over a sixteen-day period. Thoreau described their methods:

> They divided it into cakes by methods too well known to require description, and these were rapidly hauled off to an ice platform, and raised by grappling irons and block and tackle, worked by horses, onto a stack, as surely as so many barrels of flour, and these placed evenly side by side, and row upon row, as if they formed the base of an obelisk designed to pierce the clouds. They told me that in a good day they could get out a thousand tons, which was the yield of about an acre.[41]

Thoreau also wrote of the exotic destinations of his local water: "The sweltering inhabitants of Charleston and New Orleans, of Madras and Bombay and Calcutta, drink at my well ... the pure Walden water is mingled with the sacred water of the Ganges."[42]

By the 1840s, Tudor was only a small part of an expanding global ice trade, but by his death in 1864, he was still known as the "Ice King of the World," and was able to pay off all his debts.[43] Other merchants imitated Tudor in harvesting and shipping ice from New England. In 1839, New England ice ships reached Sydney, Australia, and in the 1840s, shipments to Hong Kong, the Philippines, New Zealand, the Persian Gulf, Argentina, and Peru included exported chilled vegetables, fish, butter, and eggs alongside cargos of ice. Soon, the distribution of natural ice would increase on land, as well.

Chapter 2

Mechanically Manufactured Ice

A thriving new ice market was developing in the United States, as the growing populations of New York, Philadelphia, and Baltimore were industrializing, and new ice companies sprang up there to harvest winter ice and store it for distribution and resale. In New York City, ice consumption increased from 12,000 tons in 1843 to 100,000 tons in 1856, and during the same period in Boston, consumption leapt from 6,000 tons to 85,000 tons. Due to the high volume of business, the price of ice dropped from six cents a pound to half a cent a pound. Industrial and private customers found many uses for ice and refrigeration. Fishermen began using ice to preserve their catches. Farmers used it to preserve dairy products and fresh fruit for market. Such ice was delivered by wagons to restaurants, fish and produce markets, and, of course, the private homes of the wealthy. When it arrived, more iceboxes were needed to keep it from melting too rapidly.

Inventors such as P. Kephart in 1844 (U.S. Patent 3,758) and manufacturers were now designing the familiar rectangular containers that today we would call iceboxes, but were then variously called "refrigerators," "ice boxes," "cold closets," or "fruit and vegetable preservers." They became common kitchen appliances long before the electromechanical refrigerators with which we are now familiar. They were essentially non-mechanical refrigerators, generally constructed of wood, and having hollow walls lined with tin or zinc that were filled with various insulating materials such as cork, sawdust, straw or seaweed. A large block of ice was held in a compartment near the top of the box, and the cold air circulated down and around storage compartments below to the perishable foods in the lower section.[1]

In the 1840s, household refrigerators were manufactured on the East Coast, either by Darrius Eddy of Massachusetts or Winship of Boston, and many of these were shipped west.[2] But there were functional problems.

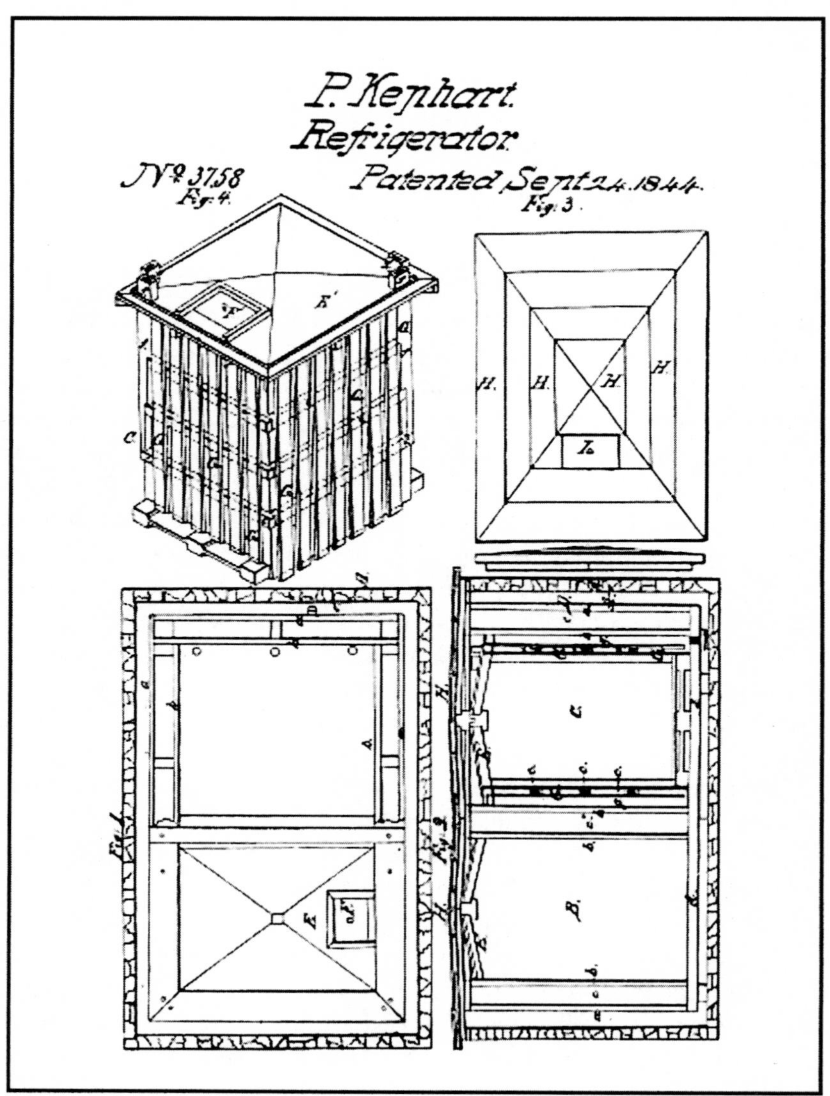

P. Kephart's U.S. Patent No. 3,758 for "refrigerator, 1844."

The early models used no moisture barrier, and condensed water on the inner metal lining rotted the wood, causing foul smells. In 1838 Henry Hill was granted U.S. Patent 750, *Improvement in Constructing Refrigerators,* to impregnate the wood liner with hot resin and beeswax. Then the inner metal lining was dispensed with, eliminating the problem. In 1845,

Thomas King, with his U.S. Patent 4,086, *Refrigerator*, dried the air in the icebox by drawing a vacuum on the box and then admitting air through a coil immersed in ice. In 1848, John Schooley devised a similar version where air flowed over ice in a separate compartment, and then flowed over into the bottom food storage chamber while rising as it was heated to a vent at the top. This idea of dry, cold air became widely employed, and refrigerators that deposited moisture on the ice, rather than on the inside of the box, were much less objectionable to users.[3]

In 1850 wealthy beneficiaries of the 1849 California gold rush created a sudden demand for luxuries, and New England companies made the first deliveries of ice by ship to San Francisco and Sacramento, along with ice-cooled apples. This established a new, lucrative, western market, but the delivery method proved to be quite expensive, and demand soon outstripped the rate of supply. In 1853, the American-Russian Commercial Company was formed to work in partnership with the Russian-American Company of Alaska (Russia owned Alaska until it was sold to the United States in 1867 for $7.2 million) to supply ice to the West Coast at $3.75 per pound ($60.90 in today's dollars). The Russian company trained Aleutian teams to harvest ice in Alaska, built sawmills to produce sawdust for insulation, and shipped ice south along with supplies of chilled fish. A competing business, the Nevada Ice Company, soon harvested ice on Pilot Creek and transported it to Sacramento at $2 ($32.50) per pound.[4]

By 1869, many middle-class homes would have iceboxes. The 1867 purchase of Alaska by Secretary of State William Seward was mocked as "Seward's Icebox." Iceboxes were described by home economist Catherine Beecher (1800–1878), who, with the help of her sister, Harriet Beecher Stowe (1811–1896, author of *Uncle Tom's Cabin* in 1852), wrote *The American Woman's Home: Or Principles of Domestic Science*, an 1869 book that proposed that a woman could run her own middle-class home efficiently on scientific principles without the servants required by homes of the wealthy. Along with detailed housekeeping procedures for all the rooms in the home, Beecher proposed and illustrated a kitchen and floor plan that provided efficient, labeled storage facilities for foodstuffs, equipment, and utensils in a compact kitchen with countertop work spaces, good window lighting and shelving. She also proposed the ideal location of sink, stove, and table to provide easy work patterns. Although there was an iron coal stove with an oven, roaster and a water heater, the refrigerator (icebox) was inconveniently located in the basement, elevated on feet and with a lining of tin filled with powdered charcoal to keep away ants. It was located there because there was usually a drain there to carry off melted ice water.[5]

But natural ice was no longer the only source of ice. As we have seen historically, man had made ice for centuries by chemical means, that is, the use of salt to lower the freezing temperature of water, and by means of rapid evaporation. In the early 19th century, experimentation had continued anonymously to produce manufactured blocks of ice by chemical means. They were made in closed metal containers, which were filled with purified water and placed into baths of brine. Pipes running through the brine carried steam, which encouraged rapid evaporation, and the water was changed to solid ice. But these were of small scale and sometimes no more than parlor tricks.[6] Producing man-made ice in quantity for commercial purposes would only be made possible by the rise of industrialization, the discovery of basic scientific principles, and the invention of complex mechanisms. By the mid–1850s, this process, called refrigeration, had reached the point of a few practical commercial applications.

The process of making ice artificially had been evolving slowly since the mid–1700s. A new area of scientific exploration had been initiated, made possible by the invention of vacuum pumps by German scientist and inventor Otto von Guericke (1602–1686) in 1650. In 1748, G. Richmann presented papers at the St. Petersburg Academy of Science noting that a wet thermometer bulb resulted in lower temperatures, and that these temperatures rose as the bulb dried. He speculated that some type of "cooling agent" was present in the atmosphere.[7] The effect of enhancing the natural evaporative cooling effect with a vacuum pump first seemed to have occurred to Scottish physician and chemist Dr. William Cullen (1710–1790), professor of medicine at the Edinburgh Medical School, Edinburgh, Scotland. In 1755, he used a mechanical pump to create a partial vacuum over a container of diethyl ether, which then boiled, absorbing heat from the surrounding air. This even created a small amount of ice, but the process found no commercial application.

However, the scientific principles Cullen demonstrated were these: (1) When any liquid turns to a vapor, it loses heat and gets colder, because the molecules of vapor need energy to move and leave the liquid. This energy comes from the liquid; the molecules left behind have less energy and so the liquid becomes colder.[8] (2) The power of evaporation of fluids is increased depending on the degree of volatility of the fluid. Diethyl ether, for example, is much more volatile than water, and thus could evaporate more readily at normal temperatures and pressures. In a vacuum the degree of evaporation is magnified even further.[9]

Most scientists in the 1700s were in Europe during America's prerevolutionary period as an English colony, Benjamin Franklin (1706–1790) was America's leading scientist of his day. His work with electricity had made him internationally famous, and in 1753 he received England's Royal Society's Copley Medal in recognition for that work. In 1756 he became one of the few 18th century Americans elected as a Fellow in the Royal Society and began to associate with British scientists. Franklin may have been aware of Cullen's recent experiments in Scotland, but he needed no such technical knowledge to notice that on a very hot day, he stayed cooler in a wet shirt in a breeze than in a dry one, a common sensation we have all experienced.

In 1758, while in Cambridge, England, on a warm day, Franklin conducted experiments to understand the principle of evaporative cooling with fellow scientist John Hadley (1731–1764), professor of chemistry at Cambridge University, by continually wetting the bulb of a thermometer with ether and using a bellows to evaporate it. With each subsequent application of ether, the thermometer registered a lower temperature, until it reached 7°F. Another thermometer in the room showed the temperature was 65°F. When the thermometer passed the freezing point of water (32°F) Franklin noted that a film of ice formed on the surface of the bulb, and when it reached 7°F, the ice was about one quarter inch thick. In his letter, *Cooling by Evaporation*, Franklin wrote: "One may see the possibility of freezing a man to death on a warm summer day."[10]

More than 30 years later, another early American inventor in Philadelphia, Oliver Evans (1755–1819), was one of the first three men to apply for patents in 1790 when the new Congress established the U.S. Patent Office. His first patent on that date was for a fully automatic gristmill to grind flour, but he also later wrote about his theories of making ice by mechanical means, using a steam engine. In his 1805 book, *The Abortion of the Young Steam Engineers Guide* (he used the term "abortion" because he was frus-

Oliver Evans (1755–1819) (courtesy Scientist & Inventors Portrait File, Archives Center, National Museum of American History, Smithsonian Institution).

trated by his inability to issue a larger and more detailed treatise on the subject). Evans noted that drawing a vacuum on water reduces its boiling point and cooled it (as proven by Cullen in 1755). He noted also that the same effect would be observed with ether, and the resulting cooling should be sufficient to produce ice (as proven by Franklin and Hadley in 1758). But as a steam engineer he went a step further to imagine a machine that actually accomplished this:

> For instance, to cool wholesome water ... for drinking ... a steam engine may work a large air pump, leaving a perfect vacuum behind it on the surface of the water at every stroke. If ether be used as a medium for conducting the heat from the water into the vacuum, the pump may force the vapor rising from the ether, into another pump to be employed to compress it into a vessel immerses in water; the heat will escape into the surrounding water, and the vapor return to ether again; which being let into the vessel in a vacuum, it may thus be used over and over repeatedly. Thus it appears possible to extract the latent heat from cold water and apply it to boil water; and to make ice in large quantities in hot countries by the power of the steam engine. I suggest these ideas merely for the consideration of those who may be disposed to investigate the principles, or wish to put them in operation.[11]

Although such a machine was never produced by Evans, his concept—evaporating (vaporizing) a liquid (called a refrigerant), which absorbed heat, then condensing it, which released heat, and repeating the process in a continuous closed cycle—would ultimately become the dominant means of mechanical cooling, which would later be described as a closed-cycle, vapor-compression refrigeration machine.[12] In such a machine, condensation of the refrigerant is done by a mechanical compressor (i.e., a pump), which squeezes the vapor into a liquid. Simply stated, the process is that a liquid removes heat from one place (a box) and as a vapor, transports it to another place (outside the box). Not a bad definition of a refrigerator.

A Massachusetts inventor and friend of Evans, Jacob Perkins (1766–1849), further developed his friend's concept. In 1816 he had worked with Evans on steam power in Philadelphia but in 1819, after Evans' death, had relocated to England to develop "un-forgeable notes" (currency) for the Royal Society. After he discovered that liquid ammonia caused a cooling effect (English scientist Michael Faraday, 1791–1867, in 1820 had just successfully liquefied ammonia by using high pressures and low temperatures), Perkins began researching refrigeration machinery and in 1834 patented, with British Patent 6,662, the first actual physical representation of Evans' 1805 concept of a closed-cycle, vapor-compression machine.

The patent was titled *Apparatus and Means for Producing Ice, and in Cooling Fluids.* Perkins briefly described his patent: "I am enabled to use volatile fluids for the purpose of producing the cooling or freezing of fluids, and yet at the same time constantly condensing such volatile fluids, and bringing them again into operation without waste."[13] Perkins persuaded a man named John Hague, a mechanic in Perkins' employ, to construct an actual machine, using ether as a refrigerant. The machine managed to produce a small quantity of ice, but it did not succeed commercially.

The purpose of refrigeration development was to chemically, or mechanically, produce a cooling effect similar to that of natural ice, and therefore, among other benefits, provide a means for preserving food. Around this time another means of preserving food was developed in France. In 1795, the French military offered a cash prize of 12,000 francs for a new method of preserving food. Nicolas Appert (1749–1841), a confectioner and chef in Paris, had been experimenting for years with placing food in glass jars, sealing them with cork and sealing wax to make them airtight, and placing them in boiling water. He submitted his invention and won the French prize in 1810. Appert's method was so simple and workable that it quickly became widespread. By 1822, the process had made its way to America, although the term "canning" would not appear until the 20th century, when mass-produced tin cans would replace the glass jars used by many to preserve summer fruits and vegetables. Before that, the process was sometimes called "appertization" in honor of Appert.[14]

Refrigeration offered a means of preserving food that surpassed all others in terms of convenience. Pickling, drying, smoking, and canning required an enormous amount of work and energy of the person doing it. Refrigeration offered the possibility of keeping food longer, more efficiently, and more effectively. To this end, John Leslie (1766–1832), professor of mathematics at Edinburgh, Scotland, had begun experimenting with absorption and vacuum refrigeration in 1810, and in 1813 published *A Short Account of Experiments and Instruments Depending on the Relations of Air to Heat and Moisture.* By 1823 he had achieved the freezing of water by evaporation and exposing it to a rarified (expanded), dry atmosphere. He placed a dish of water and a dish of sulfuric acid under a bell jar, which he evacuated with an air pump. The pump, as well as the affinity of sulfuric acid for water vapor, caused the water to evaporate rapidly and freeze. Edward Nairne in 1777 had previously used sulfuric acid to absorb water vapor to create a drier vacuum than possible with an air pump but did not use the absorption to produce cold, as Leslie did. John Leslie's absorption concept of 1823 was improved by John Vallance in 1824 with British Patents 4,884 and 5,001, in which he repeated Leslie's process more efficiently by increasing the surface area of the water with a rotating spray device for the water, increasing the surface for the sulfuric acid absorbent, and using larger pumps. Success for the absorption process of refrigeration would not be attained until Ferdinand Carré did so in 1850.[15]

Early attempts at air conditioning also began about this time in England. In December 1816, a British patent was granted to French soldier and politician Marquis Jean Frédéric de Chabannes (1770–1851) for

> a method or methods of conducting air, and regulating the temperature in houses or other buildings, and warming or cooling either air or liquids in a much more expeditious and consequently less expensive manner than hath hitherto been done.... My method of cooling air is by means of the air pump ... causing the air to pass through a cool medium.[16]

The apparatus showed a cooling tower, with which Chabannes intended to cool the air by evaporation of water. Chabannes subsequently published his proposals. At about this same time, also in England, a "Mr. Deacon" introduced an "Eolian" (meaning caused by, or carried by, the wind) heating-cooling apparatus. A fan drew air over parallel iron plates that were cooled by a cold-water bath at their base. The system was installed in some public buildings, but nothing further is known. Robert Salmon and William Warrel were issued British Patent 4,331 for an evaporative cooler in 1819. It was designed as a general means of cooling liquids, incorporating many of the elements that would be incorporated into "air washers" designed later as coolers for air conditioning. In France in 1840, Eugène Péclet would

John Leslie (1766–1832) (courtesy Scientist & Inventors Portrait File, Archives Center, National Museum of American History, Smithsonian Institution).

devise some of the earliest evaporative coolers with a modern design, but it would be another 50 years before air washers would begin to be used as air coolers. Cooling air mechanically would not take place until artificial refrigeration was invented.[17]

In about 1833, Scottish physician, chemist, inventor, and ventilation pioneer David Boswell Reid (1805–1863) generally proposed the idea of mechanical air conditioning. He wrote that it could be tested "by constructing a few chambers in every hospital, where the quality of air ... might be entirely under control, and medicated, heated, dried, moistened, cooled, and applied in any quantity, as circumstances might dictate ... in numerous cases of disease."[18] Reid constructed such a system, but not for a hospital. That idea had to wait a few years. In 1834, Reid invited participants at the meeting of the British Association, in Edinburgh, to visit his laboratory. Among those who took up his offer were some members of Parliament. Later that year, when there was a very destructive fire at the Houses of Parliament, Reid was called in as a consultant to conduct innovative work on forced ventilation. When the new building work started in 1836 for the replacement for the Houses of Parliament, the committee appointed Reid as ventilation engineer. With this appointment there began a long series of quarrels between Reid and Charles Barry, the architect for the renovations. Reluctantly, Barry eventually adopted Reid's system for the new Palace of Westminster. Reid had constructed an air-washer under the House of Lords in which the air was cleansed, cooled, and disinfected through gauze filters and water sprays. Reid described his system as follows: "Much is frequently affected in cooling the House [of Commons] in summer ... by the evaporation of water, by the contact of air with cold water apparatus (the same that is used as hot water apparatus in cold weather), and in rare cases, by the use of ice."[19] Reid's system was mechanical in the sense that that it used a fan to move the air, but it did not use a mechanical refrigeration system. Refrigeration would not be applied to strictly defined air conditioning systems until the turn of the century. However, in the 21st century, Reid would still be remembered as "Dr. Reid the ventilator."

The search for artificial ice-making continued. In 1835 a new chemical method of ice-making was accidentally discovered by French inventor Adrien-Jean-Pierre Thilorier (1790–1844), who noticed that when opening the lid of a large cylinder containing liquid carbon dioxide, most of it quickly evaporated and left only solid dry ice in the container. However, dry ice would not be produced or used commercially until 1925, 90 years later.[20] Unlike artificial water ice, it seemed to have no large-scale commercial application at the time. However, it would later become an important component of cooling technology.

All these scientific discoveries and theories contributed to a body of knowledge and principles about refrigeration, but they did not produce any practical results that had any impact on the real world. As history has shown, no modern technology arrives full-blown in an instant of time, created by any one individual; rather, it evolves over a long period of time in bits and pieces as it is refined by generations of individuals. What was needed commercially in the technology of mechanical refrigeration was the mass production of mechanically-made water ice, to provide a reliable, year-round source of ice not dependent on the seasons or weather; and machinery to provide reliable and effective comfort cooling for buildings. The first pragmatic step in this direction came from an unexpected source, the medical field.

American physician and inventor Dr. John Gorrie (1803–1855) had moved in 1833 to Apalachicola, Florida, a small, coastal cotton port on the Gulf of Mexico at the mouth of the Apalachicola River. He was soon intently involved with town affairs. In 1834 he was made postmaster and in 1836 became president of the Pensacola Bank. That year, the Apalachicola Company asked him to report on the effects of climate on the population, as they considered expansion of the town. Gorrie accepted the common miasmatic theory of the time that it was "bad air" that caused disease. Diseases, including yellow fever, cholera, and typhoid fever, were thought from ancient times to be caused by impure, noxious air emanating from rotting organic matter in swamps, which was also called "miasma," or "night air." Based on this knowledge, Gorrie recommended draining swamps in the low-lying areas of the town, and recommended that only brick buildings be constructed.

Dr. John Gorrie (1803–1855) (courtesy Bureau of Operational Services, Florida Department of Environmental Protection).

By 1837, Apalachicola had grown to a population of 1,500 with increased trade in cotton, and Gorrie, a prominent community physician, suggested that the town needed a hospital. There was already a small medical unit in operation at a local U.S. Marine Hospital, and Gorrie was a part-time employee there, where he studied tropical diseases. Most of his patients were sailors and waterfront workers, and most of them had fever of some kind every summer in Apalachicola. He became obsessed with finding a cure for the diseases, and had come unknowingly close to doing so in 1836, over 60 years ahead of the rest of the world, when he wrote: "Gauze curtains, though chiefly used to prevent annoyance and suffering from mosquitoes, are thought also to be sifters of the atmosphere and interceptors and decomposers of malaria."[21] Although Gorrie clearly suspected that the air, not mosquitoes, caused disease, even such a mention of the mosquito connected to disease, let alone that it might be the actual disease carrier, would not be made until 1881, long after Gorrie's death. By 1838, Gorrie did notice that malaria seemed to be connected with hot, humid weather, and accordingly, he looked for ways to lower the temperature of his patients in the summer.[22] To accomplish this, he placed ice in a basin suspended from the ceiling and blew air over it with a fan. Heavier cool air flowed down across the patient and cooled the air considerably. But large amounts of ice were needed on a continuing basis, and ice had to be brought by boat from the northern lakes. (Tudor's ice boats did not often reach small ports like Apalachicola, and even if they did, the price was an exorbitant $1.25 per pound, or $32.30 in today's dollars.)

In 1842, Gorrie, in an effort to alleviate these prohibitive costs, began to explore the possibility of using mechanical refrigeration, primarily for humanitarian reasons. At first, he idealistically envisioned the largest benefit to mankind he could imagine. In an anonymous article, he proposed that cities should be cooled by artificial means, "to counteract the evils of high temperature, and improve the condition of our cities ... [by] the rarification [expansion] and distribution of atmospheric air, previously deprived of large portions of latent caloric by mechanical condensation."[23] Essentially, Gorrie was the first to envision what we now call air conditioning. Gorrie described how this might be accomplished:

> Whenever the escape of air ... takes place, it will expand, and in the process, precisely the quantity of heat which was previously obtained from it will be absorbed from all surrounding substances, and rendered latent. Acting on this powerful source of heat, by means of water, wind, or steam power, into suitable reservoirs in the suburbs of cities, and thence to transmit it through conduits, like water or gas, so that it may

> be distributed to, and set free in the houses, and even in the streets and squares of the city.[24]

His article then discussed the "condensation" (compression) and the "rarification" (expansion) of air, and the advantages of cool, dry air. Over the next two years, Gorrie constructed a working air-cycle refrigeration system, which he described in a series of articles in 1844 for the *Commercial Advertiser,* a local Apalachicola newspaper under the nom-de-plume "Jenner." Leaving out his earlier plans to cool and ventilate entire cities, he concentrated on the problem of humans and diseases:

> Let the houses of warm countries be built with an equal regard to insulation, and a like labor and expense be incurred in moderating the temperature, and lessening the moisture in the internal atmosphere, and the occupants would incur little or no risk from malaria.... Atmospherical temperature determines, or at least greatly modifies the character of our race.[25]

Gorrie then described his idea for "an engine for ventilation, and cooling air in tropical climates," which compressed air with a double-acting piston pump, forcing it into a storage tank, through a "weighted valve," then into a double-acting "expansion engine." This engine was connected to the compressor so as to "exert the same mechanical force that was required" to compress the air. Horse, water, or steam power could power the compressor.[26]

The principle that compressed air could provide cooling when it performed work was observed in the eighteenth century by Evans and Perkins but not pursued practically until Gorrie's work. His system used the principle observed by John Dalton, a professor at New College, Manchester, England, who conducted experiments in 1802 and noticed that the cooling effect upon air depended on whether it was rarified (expanded) or compressed.[27] Later, the work of James Prescott Joule (1818–1889) and William Thompson (1824–1907) proved that when air is forced through a valve while being kept insulated, so that no heat is exchanged with the environment, the air is cooled. In thermodynamics, this is called the Joule-Thompson effect. Such refrigeration systems based on this principle are known as closed "air cycle" machines.

In 1845, Gorrie gave up his medical practice to perfect these systems. His main intention was to use mechanical refrigeration for humanitarian reasons, from cooling sickrooms to entire houses.[28] But he soon realized that the only possible way to realize his humanitarian dreams would be to commercialize his refrigeration machine, and the only possible commercial application for such machines in the 1840s was ice-making, so his work shifted to that goal.

Gorrie's system was announced at a gala celebration of Bastille Day on July 14, 1850, at the Mansion House Hotel in Apalachicola, where the French consul in Apalachicola, Monsieur Rosan, was celebrating the occasion. No ice ship had arrived, so the champagne was to be served warm. But at the moment before the toast to the French Republic, four servants entered, each with a silver tray on which was a block of ice the size of a house brick, to chill the wine, as one guest put it, "by American genius." They had been made by Gorrie's prototype ice machine. The *New York Times*, however, was skeptical, to say the least, at this news. It reported: "There is a crank down in Apalachicola, Florida, who claims that he can make ice as good as God Almighty."[29] Gorrie's machine used a steam engine to compress air and then partly cool the hot compressed air with water before allowing it to expand while doing part of the work needed to drive the air compressor. The expansion of the air cooled enough to freeze water and produce ice, or, as his patent specifically stated, to flow "through a pipe for effecting refrigeration otherwise."[30] In other words, it could also function as an air-conditioning system, a potential application that could realize his original dream of cooling hospitals.

Gorrie advertised his invention as "the first commercial machine to work for ice making and refrigeration.... A ton of ice can be made on any part of the earth for less than $2" (one hundredth of a cent per pound, incredibly cheaper than the $1.25 per pound for shipped ice). Gorrie obtained some capital from a New Orleans businessman and attempted to raise financial backing from a Boston man, but they both failed to come through, and commercial development became impossible.[31] However, Gorrie proceeded to file for patents in 1849, and in 1850 he was awarded British Patent 13,234 and a U.S. Patent in 1851 (U.S. Patent No. 8,080, *Improved Process for the Artificial Production of Ice*) for his "ice machine."

Gorrie built a working prototype, but the system unfortunately leaked, performed only irregularly, and was a commercial failure. He next attempted to license his machine using the services of a New York agent, and published a pamphlet in 1854 that described and illustrated a much more refined ice-making machine. But there is no evidence that his machine was ever licensed or produced. Gorrie's extreme efforts and disappointments took a toll on his health and he passed away, a broken and dispirited man, in 1855.[32] He is buried in Gorrie Square, Apalachicola, with a statue honoring him. Gorrie blamed Frederic Tudor, the "Ice King," for his failure, suspecting that Tudor had launched a smear campaign against his invention.[33] Nevertheless, he was the first to conceive of a

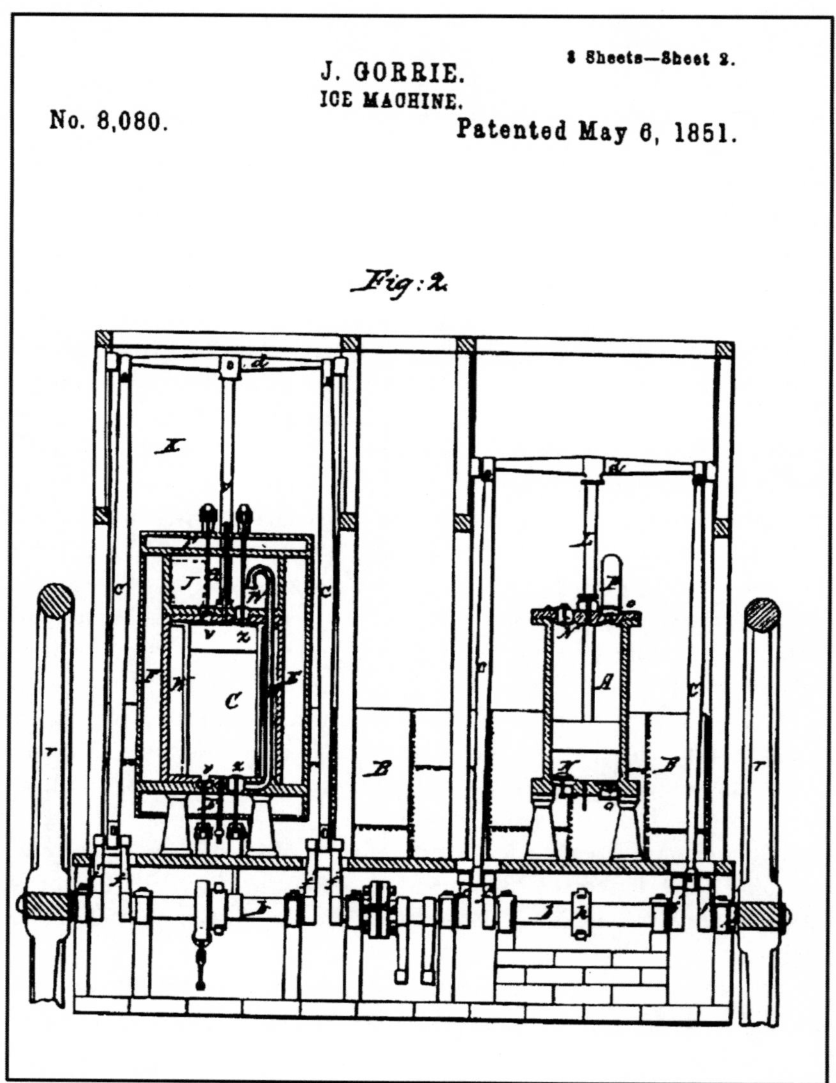
J. GORRIE.
ICE MACHINE.
No. 8,080.
Patented May 6, 1851.
3 Sheets—Sheet 2.

Gorrie's U.S. Patent No. 8,080, 1851.

system, air conditioning, to cool the air for comfort in homes and hospitals.

He was not the only one thinking along those lines, however. In the 1853–1854 *Proceedings of the Philosophical Society of Glasgow* 3, William Thompson, Irish and British mathematical physicist and engineer at the

University of Glasgow, deep in the study of thermodynamics, had written a treatise: *On the Economy of the Heating and Cooling of Buildings by Means of Currents of Air*.[34] In 1892, Queen Victoria would knight Thompson for his work on the transatlantic telegraph in the 1860s as well as his work on thermodynamics, and he became Sir William Thompson. He would then adopt the Irish title Baron Kelvin, and therefore often is called Lord Kelvin. He is famous for his determination in the 1840s of the correct value of absolute zero—approximately –273.15° Celsius. Absolute temperatures today are stated in units of kelvin, in his honor.

Independently, Charles Piazzi Smyth (1819–1900), astronomer royal in Scotland, also had similar aspirations to Gorrie's in using mechanical refrigeration for comfort cooling. In 1850, he described his experiments toward this objective starting in 1845, and by 1847 had built an apparatus to test his theory. He went on to describe his proposed machine, powered by oxen, to be used to provide cool air in the tropics. His concept, also based on the compression of air, used expansion of air through weighted valves but later included an expansion cylinder. His creative ideas aroused interest in the scientific community, and beginning in the 1860s, this approach was intensively developed in Great Britain.[35] Coincidentally, three of these early developers in air conditioning—Gorrie, Reid, and Smythe—all specifically intended them for hospitals.

Numerous other patents for ice-making machines were filed by this time, based on cooling effects that depended on the expansion of compressed air or upon the evaporation of very volatile liquids such as liquefied ammonia. But as we have already seen, patents are one thing—turning them into commercial success is another. One of those who did succeed was American engineer, scientist and inventor Alexander Catlin Twining (1801–1884) of Cleveland, Ohio, a civil engineer, who had been engaged in the construction of a number of railroads and canals in Ohio, but had abandoned that profession to devote his life to scientific research. He began experimenting with vapor-compression refrigeration in 1848 and obtained British and American patents in 1850 and 1853 (U.S. Patent 10,221, *Manufacturing Ice*). He described his early experiments:

> By maintaining a vacuum in a small reservoir of ether immersed in water, the weight of ice which the evaporation of a given quantity of ether would produced, was proven.... The next question arising was whether the ether vapor could be re-condensed with sufficient rapidity. By numerous experiments it was ascertained that only 200 superficial feet of thin copper pipes would form an adequate surface for the manufacture of 2000 lbs. of ice in a day ... even employing water of the temperature of the Earth's equator....

> The first attempt at a complete freezing machine was made in the summer of 1850. The machine had only capacity to freeze a pail of water at one operation ... but became an encouragement to attempt a vastly larger construction.[36]

Such a larger machine was actually constructed at the Cuyahoga Steam Furnace Company in Cleveland, Ohio, with freezing trials beginning in February 1855. By that summer, the machine had produced 1,700 pounds of ice; it continued in operation over the next two years.[37] Twining is credited with the initiation of commercial refrigeration in the United States in 1856.[38] Elements of his patents were incorporated into the first commercial ice plant in the southern United States at the Louisiana Ice Manufacturing Company in 1862.

Twining attempted to interest financial backers for construction of an ice plant in New Orleans, but the start of the Civil War discouraged any progress in this venture. In 1863 he tried to set up an improved system at the Morgan Iron Works of New York City, but a shortage of funds, accidents, and construction imperfections resulted in losses to Twining of more than $36,000 ($689,000 in today's dollars). Twining felt that the Civil War had not only prevented his efforts to succeed, but had given others, such as Ferdinand Carré and James Harrison, the opportunity to steal his ideas for their own use.[39]

Scottish engineer and British journalist James Harrison (1816–1893) had emigrated from Glasgow to Australia in 1837 and became editor and sole owner of the *Geelong Advertiser* by 1842. While cleaning movable type with ether, he suddenly noticed that the ether would leave the metal type extremely cold to the touch. He made an improved ether-compression machine and began operations in 1851 in Geelong, Australia.[40] Harrison and his blacksmith friend and machinist, John Scott, built the first commercial machine in 1854 and his patent was granted by the Colony of Victoria, Australia in 1855.[41] He then went to London where he patented the process (British Patent 747 of 1856) and also patented his apparatus (British Patent 2,362 of 1857).[42] The latter patent was a new, more effective design prepared in close collaboration with Daniel Siebe of the British steam engineering company Siebe and Company, during the time while Harrison was in England for two years. The patent covered a simple vapor-compression machine using ether for a refrigerant and a compressor to force the refrigeration gas to pass through a condenser, where it cooled down and liquefied. The liquefied gas then circulated through the refrigeration coils and vaporized again, cooling down the surrounding system.

The first machine of this improved design, built by Siebe, was sold

Harrison-Siebe machine, 1857.

to the Truman, Hanbury, and Buxton Brewery in London. The machine used a 16-foot flywheel and produced an astounding 6,000 pounds of ice per day. It was used in 1859 to establish the Victoria Ice Works in Australia.[43] Another machine built by Siebe, with a capacity of 8,000 lbs. per day, was exhibited in London before being exported to Australia after Harrison's return there in 1858 to promote the use of his machines.[44] Thus Harrison was the first refrigeration pioneer to see actual commercial production of his invention, made possible by the expertise of a steam-engineering firm. His machine was provided to an Australian brewery, as well as the one in London, making it possible to brew lager beer even in hot weather.[45]

Natural ice found an unusual application in 1859, after the Comstock Lode was discovered in Nevada. It was the first major discovery of silver in the United States and the silver rush drew thousands of prospectors to Mount Davidson near Virginia City to mine gold and silver. As the miners dug deeper and deeper (eventually to 3,000 feet), the temperature in the mines continued to rise, reaching well over 110°F. They started having to give the miners ice to help them survive. A miner could use 75 pounds of ice per shift. The ice was harvested in winter from frozen lakes in the Sierra Nevada. Peak production of the mines was in 1877, and it tapered off after 1880.

Gorrie and Smyth's experimentation with compressed-air (air cycle) systems had never been commercially developed, but when Scottish engineer Alexander Carnegie Kirk (1830–1892) was instructed to find a replacement for one of Harrison's vapor-compression machines, because of the fear of using its flammable ether refrigerant, he developed successful closed-cycle, air-compression refrigeration machines before 1862, some of which operated quite satisfactorily for ten years. In Brunswick, Germany, engineer Franz Windhausen (d. 1904) would patent his air cycle machine in 1869, and about a hundred of these were sold but were prone to periodic clogging by snow. Paul Giffard of France tried for years to solve this problem and developed an open-cycle, cold-air machine in 1870 but was not successful in totally overcoming the clogging problem.

The first air-cycle machine to see long-term operation was a British machine, the Bell-Coleman. Henry and Joseph Bell of Glasgow, Scotland, wanted to send fresh meat by ship and consulted with William Thompson (Lord Kelvin), who directed them to a chemist, Joseph Coleman. Coleman studied the Giffard machine but found no suitable machines after a search of one hundred patents, so Coleman and the Bells produced a machine of their own design in 1877 under the name of the Bell-Coleman Mechanical Refrigeration Company. Within two years, their machines were installed on the steamships *Circassia,* for trips to New York, and *Strathleven,* traveling between England and Australia.[46] Air-cycle machines were popular from the 1880s, in mostly British firms, but would be displaced after the turn of the century by smaller, more efficient carbon dioxide vapor-compression machines.[47] In 1878, Alfred Seale Haslam (1844–1927) of England bought the Bell-Coleman patents and would eventually equip over four hundred plants and ships with Bell-Coleman machines, including the liner *Orient* in 1881.

By the 1860s, after Harrison's machine had been used in an Australian brewery, natural and manufactured ice was being used regularly to allow the year-round brewing of the highly popular lager beer. Until 1860, the most popular beer was ale, which did not require cooling. An English beverage since the Middle Ages, ale was, and still is, usually drunk at room temperature by the British.[48] It is brewed using yeast, which ferments on the surface of the beer vat, over a period of 5 to 7 days, at a temperature ranging from 60° to 70°F. On the other hand, German lager beer is a type of beer that is fermented and conditioned only at low temperatures. Lager beer is produced by yeasts that ferment on the bottom of the vat, and the process takes place over a period of up to 12 weeks, in an ambient temperature of just above the freezing point (32°F). During this time, the beer was stored in cold cellars. Cold storage of beer in caves (*lagering* in German, from *lagern,* to store) was common throughout the Middle Ages. To brew at higher temperatures was very likely to cause bad beer.[49] In Germany in the 19th century, however, lager brewing had been banned during the summer, due to the lack of large-scale cooling facilities. But by 1870, breweries were the largest users of both natural ice and commercial refrigeration. By then the number of ale breweries would fall from 281 in 1860 to 18 in 1870, while those brewing lager would increase from 135 in 1860 to 831 in 1870, primarily because the use of ice and the development of ice-making machines enabled the year-round extension of the lager brewing season.[50]

Elsewhere in the world, mechanical refrigeration was being refined to mass production level for the export of frozen meat. Australian industrialist Thomas Sutcliffe Mort (1816–1878) had a large cattle estate on the coast of New South Wales, where he produced milk, cheese, and butter for the Sydney market. Looking at refrigeration to improve his operations, he financed experiments by French-born engineer Eugene Dominic Nicolle, who had arrived in Australia in 1853 and registered his first ice-making patent in 1861. Mort eventually would establish what is said to have been the first cold storage, meat-freezing works using mechanical refrigeration in the world at Darling Harbour, Australia, which began operations in 1875 and later would became the New South Wales Fresh Food and Ice Company.[51] The complex would have a central operations with gauges, control valves, and remote-reading thermometers. Nicolle would employ an air-to-air heat exchanger with a ventilation supply and exhaust to improve energy efficiency.[52]

Meanwhile, in France, the brother of engineer Ferdinand Carré (1824–1900), Edmond, had developed the first absorption-process refrigerator in 1850, using water and sulfuric acid. Ferdinand had also experimented since 1836 with an ether vapor-compression process, but he switched to the absorption process when he continued his brother's work and in 1858 developed a machine that used gaseous ammonia dissolved in water as a refrigerant (referred to as "aqua ammonia"). It was patented in France in 1859 and in the U.S. in 1860 (U.S. Patent 30,201, *Improvement in Apparatus for Freezing Liquids*).[53] It was of a small size, suitable for households, and would see extended usage for many years. He would exhibit it at the Universal London Exhibition in 1862, where it would produce 440 pounds of ice hourly.[54]

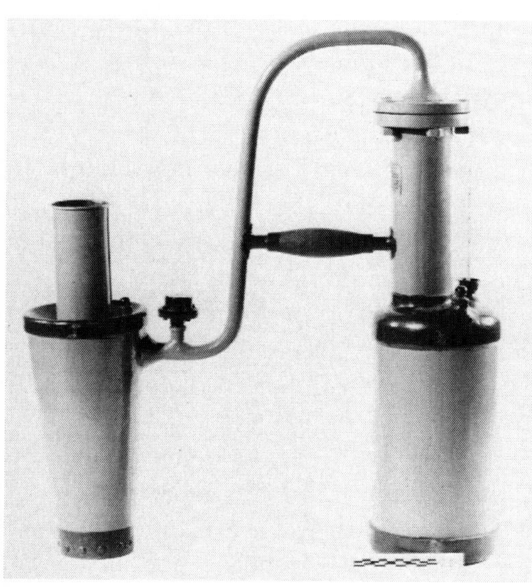

Carré absorption device, 1859 (courtesy the Division of Work and Industry, National Museum of American History, Smithsonian Institution).

The absorption process is quite different than the compression process. A refrigerant (usually ammonia) is heated by a gas flame to cause it to vaporize, which causes the ammonia gas to be absorbed into a liquid (water). As it is absorbed, it simultaneously cools and condenses, and automatically alters the pressure in the closed system, keeping the refrigerant flowing, thus making it possible for heat to be absorbed in one place (the box) and released in another (outside the box), just as the compression process does. The absorption system would become an alternative to the compression system, but it had a major advantage—it was quiet, whereas mechanical compressors were quite noisy. At the same time, Carré also developed a continuous absorption machine for commercial use, manufactured in Paris by Mignon and Rouert in 1861.

It is not surprising that mechanical ice production in the United States started in the South, where the summers were hotter and discomfort from the heat affected a larger population than in the North. During the American Civil War, the natural ice supply from the North was cut off to the South, but ingenious men in Texas shipped a Carré machine through the Union blockade into Mexico and eventually to Texas, where it was put into operation in San Antonio. Around 1865, Daniel Livingston Holden (1837–1924), an engineer with Mepes, Holden and Montgomery and Co., who had supervised the installation of the Carré machine, made several improvements to the machine, equipping it with steam coils and using distilled water to make clear ice. In 1866, he secured the rights from Peter van der Weyde to use a spirit called "chimogene" (a petroleum ether) as

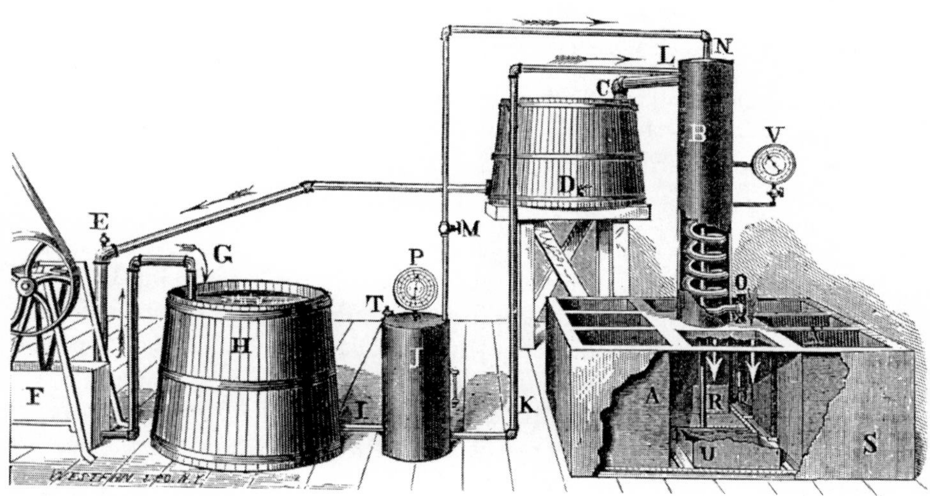

Holden ice machine, 1869. B is heat exchanger; D is water to be frozen; U is evaporator ("Making Ice by Machinery,"* The Manufacture and Builder, *vol. 1, December 1869, 353).

a refrigerant, and began the building of ice-making plants. By 1867, three companies would be manufacturing ice in San Antonio, when there were only five in all of the United States. Holden supervised the installation of a 60-ton-capacity plant in New Orleans and in 1869, would take out a patent on his own designs (U.S. Patent 95,347, *Improvement in Ice Machines*). By 1878, Holden's ice plants were being built at the Penn Iron Works in Philadelphia.[55]

During the Civil War, soldiers recovering from wounds would be served ice cream occasionally at the large general hospitals located in big cities such as Philadelphia, New York, Washington, D.C., and Chicago. The well-known poet Walt Whitman served for a time at one of those hospitals, and it is documented that on more than one occasion, he bought ice cream for the men on his ward out of his own pocket. Northern troops occupying Southern regions had a difficult time dealing with the heat and were desperate for cool refreshments. A Scottish immigrant in Demopolis, Alabama, David Boyle (1837–1891), took full advantage of this to start a business in 1865. He described his initial inspiration:

> I was keeping store and making and selling ice cream and lemonaid [*sic*]. A brigade of Federal troops was stationed there, and it were [*sic*] a bonanza to me. I had a shipment of ice from New Orleans delayed in transit three or four days, and when it reached Demopolis its actual cost was

> about seventy-five cents [$11.10 today] per pound. The weather was hot, and it did not take long to get rid if it. I used it to cool lemonaid, and sold it at a good profit to the Yankee soldiers. The unreliability of transportation, the high cost, and the absolute need of ice at Demopolis set me to thinking and determined me to attempt the making of a machine to supply the wants of Demopolis. Just think of it! The wants of Demopolis! And that was my idea.[56]

Boyle heard of an ice machine in New Orleans (probably Carré's) and went there, only to find it was too expensive for commercial use. But he sold his assets and bought a machine (probably Holden's), which turned out to be a total failure. In 1869 he took his family to San Francisco and spent a year in the Mechanics' Institute there, learning about ice machines. Frustrated by attempts to find a satisfactory machine made by others, he moved back to New Orleans and constructed a one-ton machine, which Boyle himself said "leaked like a sieve" and had to be reconstructed. He would continue to improve it (U.S. Patents 163,142 and 163,143) and by 1877 would become a successful manufacturer.[57]

In 1867, French immigrant Andrew Muhl of San Antonio, in partnership with a man named Paggi,[58] had built an ice-making machine (U.S. Patent 121,402, *Improvement in Apparatus for the Manufacture of Ice*) in San Antonio to help service the expanding beef industry before moving it to Waco in 1871. In 1873 the patent was contracted by the Columbia Iron Works, a company acquired by the W.C. Bradley Company, which went on to make the first U.S. commercial ice-making machines.[59] Back in France, meat refrigeration was also being explored by French engineer Charles Tellier (1828–1913), called by the French the "Father of Cold," who had constructed a vapor-compression machine using methyl ether in 1868 (U.S. Patent 85,719, *Improved the Manufacture of Ice and the Refrigerating of Air Liquids &c*). That year, he tried to ship refrigerated meat to London aboard the *City of Rio de Janeiro*, but his equipment failed and the meat spoiled. He then established a plant to produce ice and "carafes frappes" (a frozen dessert made with shaved ice) in Marseilles in 1869, but it was not a commercial success. He next built what was possibly the first mechanically refrigerated cold-storage plant in Auteuil, France, and was granted U.S. Patent 85,719 for it in 1869, but the plant also failed due to the outbreak of the French-German War in 1870. Nevertheless, his unique use of forced refrigerated air as a cooling means was an early demonstration of air conditioning.[60]

Between 1871 and 1881, the first mechanically refrigerated abattoir (French for "slaughter house") in the United States was planned, established, and successfully operated in Fulton, Texas, for the purpose of chilling and curing beef for shipment to Liverpool, England, and other destinations. Daniel Livingston Holden, his brother Elbridge, and Elbridge's father-in-law George W. Fulton took part in this new process of beef packing and shipping. Thomas L. Rankin, of Dallas and Denison, held many patents in refrigeration and had been involved in work with Daniel Holden. From 1870 to 1877, Rankin worked on the development of refrigerator and abattoir service for rail shipping of refrigerated beef from Texas and the Great Plains. In late 1873, the Texas and Atlantic Refrigeration Company of Denison made the first shipment of chilled beef across the country from Texas to New York. The development made by Rankin and his Texas associates spread rapidly to other beef shipping centers of the nation.[61]

Many people were trying to figure out how to air-condition spaces with the technology at hand. The 1860s brought a more sophisticated level of experimentation to the comfort cooling of rooms. In 1864, George Knight of Cincinnati proposed a hospital cooling system, described in *Scientific American,* composed of an air washer to clean the air, a fan-coil supplied by melting, salted ice to cool the air, and an overhead distribution system with individually dampered outlets. Even politicians were motivated to devise some sort of comfort cooling. Daniel E. Somes (1815–1888) of Washington, D.C., was a U.S. representative from Maine in the 36th Congress (1859–1861) who proposed that "hospitals may be so arranged that heat, flies, nor dust, need ever be present to harass and torment the patients. This is accomplished also at a cost so comparatively small as scarcely to deserve a mention. Had Mr. Somes made no discovery but this, he would be entitled to (and receive) the gratitude of the (human) race."[62]

As a patent lawyer after his term, Somes obtained numerous patents for comfort cooling ideas during the period 1864–1869, including U.S. Patent 44,229, *Improved Means of Securing a Uniform Temperature in Packing and Preserving Houses, Hospitals, and Other Buildings*; U.S. Patent 46,596, *Improved Mode of Cooling and Ventilating Dwellings, Churches, Hospitals, Theaters, and Other Buildings*; U.S. Patent 61,886, *Improvement in Moistening, Cooling, and Warming Air*; U.S. Patent 70,909, *Improvement in Cooling Air and Other Substances*; U.S. Patent 73,936, *Improved Apparatus for Making Ice, and for Cooling Air and Liquids*; and U.S. Patent 77,669, *Improvement in Cooling Air and Liquids in Making Ice.* Some of these proposed cooling buildings by spraying mists on building roofs and walls, or bringing in air that was cooled by underground ducts in which cold water was circulated to them.[63] Nathaniel Shaler of Newport, Kentucky was granted U.S. Patent 47,991 in 1865 for an *Improved Air Cooling*

Apparatus that was a heat exchanger made with "ice holders" placed in a "tortuous passage" through which room air is blown to cool it.[64]

Numerous patents would be issued after the 1870s for comfort cooling schemes. In France in 1873, A. Jouglet wrote a treatise on comfort cooling, *On the Various Systems of Cooling the Air*, in which he discussed

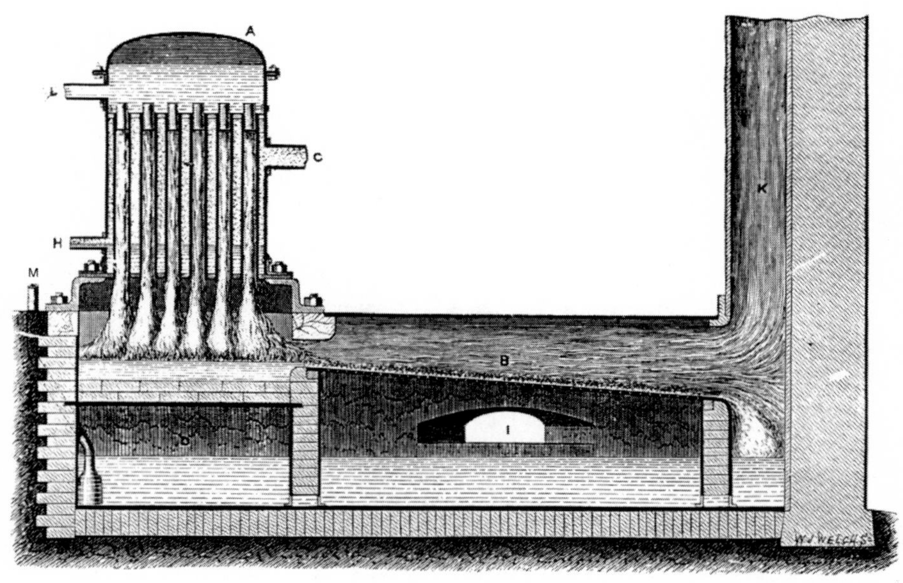

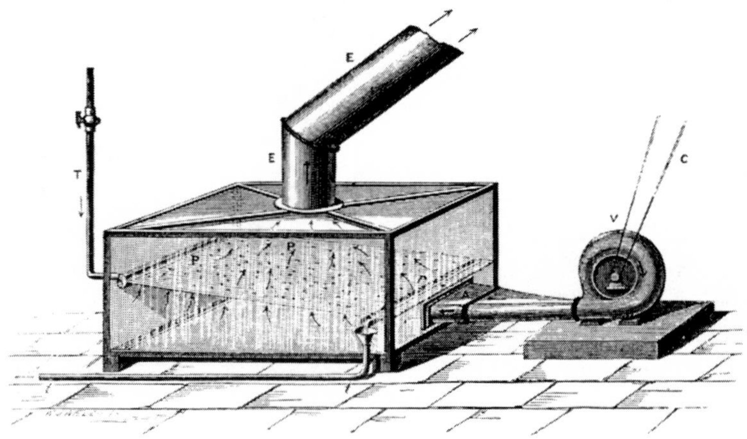

Evaporative air cooler and air washer of the 1870s, by A. Jouglet (*The Practical Magazine*, 1873, 459).

different methods of cooling air, including compression-re-expansion of air, cooling by evaporation, cooling by ice or refrigeration machines, and cooling by circulation of air through underground ducts.[65] However, regarding cooling with ice, he concluded his treatise disappointedly: "In point of fact, this method of refrigeration must be considered as impractical, while ice is not very cheap, and cold cannot be produced as inexpensively as heat."[66]

"Air conditioning" at this time, however, was still possible mainly with the traditional use of fans to move air and increase the evaporation of perspiration. Handheld fans were still in common individual use, but ceiling fans with two blades were becoming mechanized by placing turbines in nearby running water sources, which powered multiple ceiling fans in series, all driven by a leather belt on pulleys, which connected to the rotating turbine outside. In this way, large rooms, such as in hotels, meeting places, or churches, could be cooled, more or less automatically. Inventors were also thinking about smaller, more personal systems of cooling, as well. In 1857, U.S. Patent 18,696 by David Kahnweiler proposed a *Ventilating Rocking Chair,* which, when rocked, activated a bellows beneath the chair, which blew air through a Rube Goldberg assembly of pipes through an icebox, and directed the cooled air through a nozzle, which was pivoted back and forth by the arm of the rocker, into the face of the user.[67]

Meanwhile, the natural ice trade was experiencing serious difficulties. In 1860 there occurred the first of four disastrous ice-famines during warm winters that prevented the formation of ice in New England, creating severe shortages and driving up the prices of ice. As a result, James L. Cheeseman, who had a successful ice harvesting business on the Hudson River, moved his operations to Maine, where winters were colder. Although the Civil War disrupted ice delivery to the South, it was Maine merchants who supplied the Union Army, which used ice in its Southern campaigns.[68]

The United States was also expanding westward with the railroads in the 1850 and 1860s. Special refrigerator railroad cars were developed, such as one patented by J.B. Sutherland in 1867 (U.S. Patent 71,423). The railroads would soon become the "ice ships" of the country's interior. Hiram Joy began exploiting ice in Crystal Lake, near Chicago, which was linked to the city by the Chicago, St. Paul, and Fond du Lac Railroad. Both Chicago and Cincinnati, Ohio, began to use ice in packing pork for rail shipments, and chilled meat shipments, multiples each day, were being sent from Chicago to the East. Chilled butter from the Midwest was

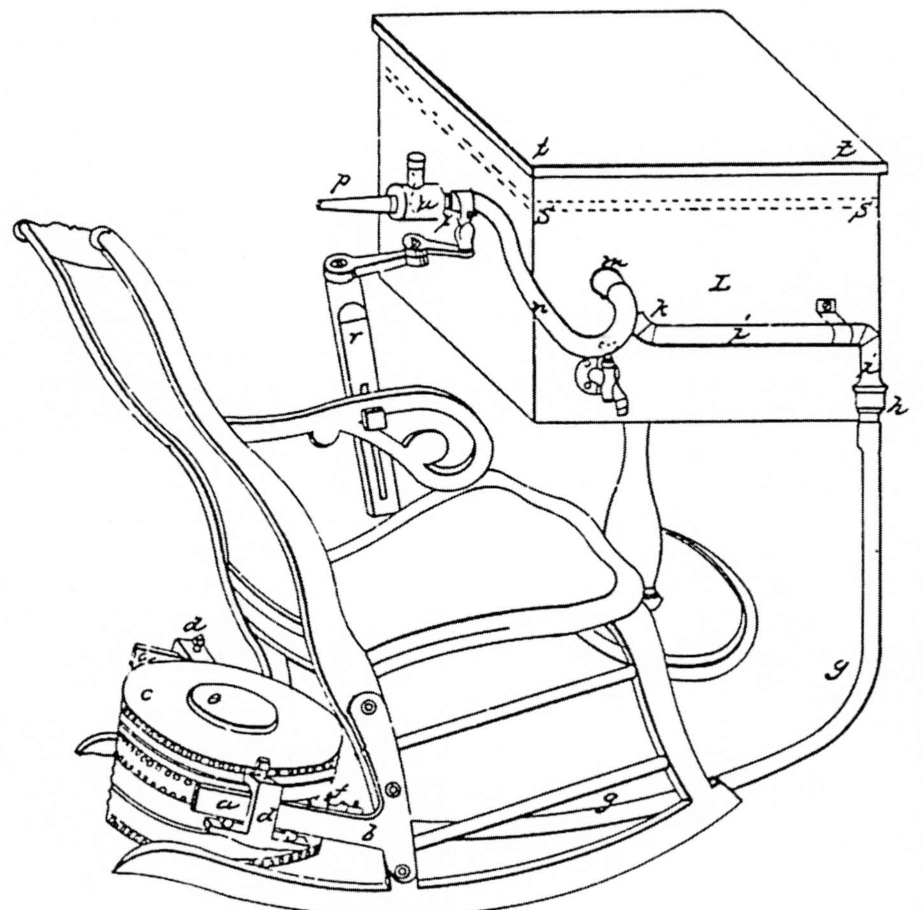

Kahnweiler "ventilating rocking chair," 1859, U.S. Patent No. 18,696.

shipped to New York, and then onward to Europe. A chain of ice stations in Chicago, Omaha, Utah, and the Sierra Nevada permitted railroad refrigerator cars to cross the continent by renewing their supplies of ice. The ability of ice companies to ship ice across the country would end the Alaska ice trade, which disappeared in the 1870s and 1880s due to the competition by rail from the East. The rival Chicago-based meat firms of Armour and Swift would enter the refrigerated meat transport market in the late 1870s by establishing their own fleet of ice-cooled refrigerator cars designed by Henry A. Roberts for Gustafus F. Swift in 1877 (U.S. Patent 256,052) and by establishing a network of icing stations across the country. As a result,

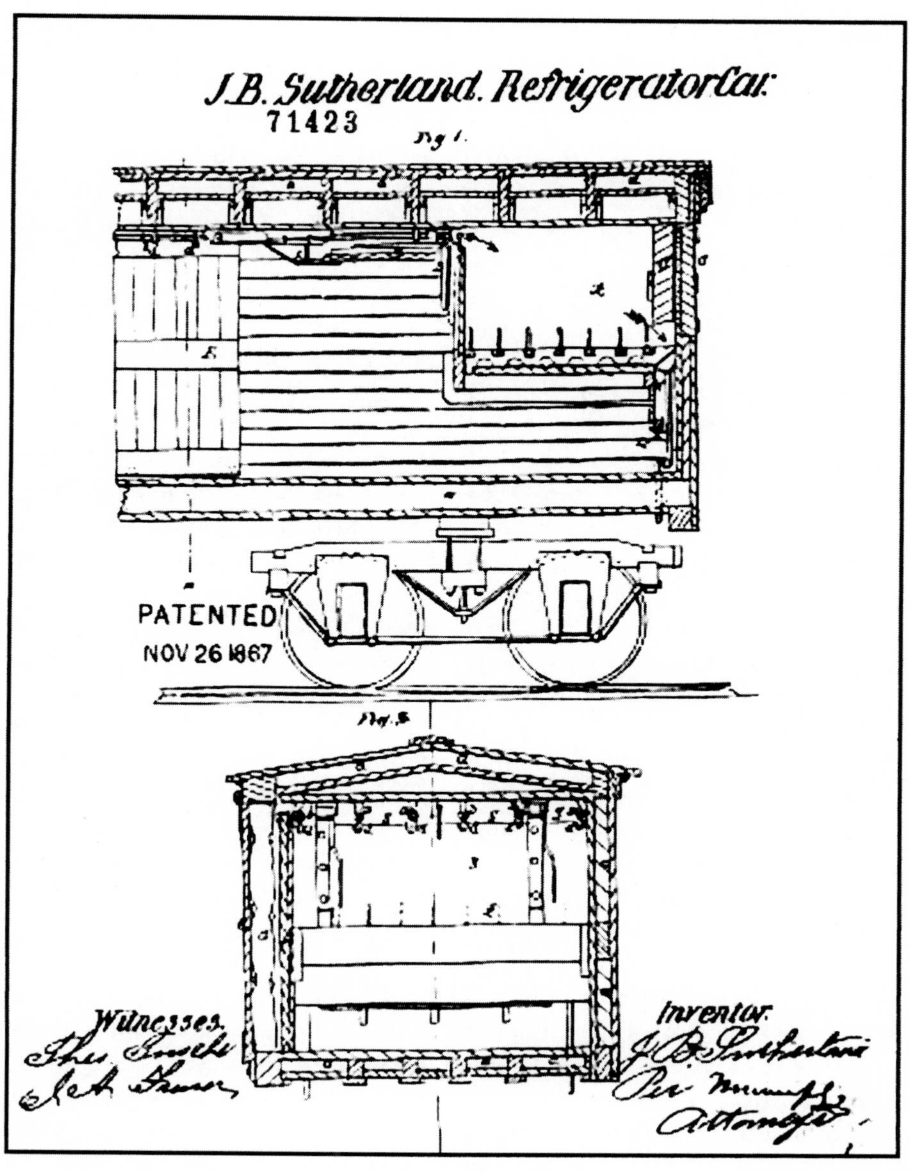

J.B. Sutherland's railroad refrigerator car, U.S. Patent No. 71,423, 1867.

sales of chilled Chicago beef to the eastern seaboard would increase from 15,680 tons a year in 1880 to 173,067 tons in 1884.[69]

Natural ice exports abroad from the United States peaked around 1870, when 65,802 tons worth $267,702 ($4,420,000 in today's dollars) were shipped from U.S. ports. During this time, ice began to be used by

Timothy Eastman, of the Bell Brothers firm, to ship American meat to Britain in 1875. The following year, 9,888 tons were shipped by Bell. The chilled meat was retailed through special warehouses and stores in Britain. By then, the ice trade was well developed in Europe, with British interests importing ice from Norway starting in the 1850s, a business that would peak during the 1890s.

In the early 1860s, Lake Oppegård in Norway had been re-named "Wenham Lake" to confuse its product with the genuine, crystal-clear New England "Wenham Lake" specialty. A major British company, Leftwich, which imported the most ice from Norway, would keep a thousand tons of ice in storage at all times to meet demand in 1900. The Norway business would later be taken over by Norwegians themselves. Austria entered the European ice market after Norway, and the Vienna Ice Company exported natural ice to Germany by the end of the century. By the 1870s, hundreds of men were employed to cut ice from the glaciers at Grindenwald in Switzerland.[70]

Experimentation with ice-making machines and refrigerated shipping continued, sometimes by well-known personalities. President Abraham Lincoln appointed Thaddeus S.C. Lowe (1832–1913), a prominent U.S. scientist and inventor, as chief aeronaut of the new Federal Army Balloon Corps during the Civil War. While experimenting with carbon dioxide for military balloons, Lowe realized the possibility of using it as a refrigerant, and he applied for and held numerous patents on ice-making machines. His "Compression Ice Machine" would revolutionize the cold-storage industry. He was granted British Patent 952 in 1867 (and U.S. Patent 68,413, *Improved Mode of Manufacturing Ice*) and built an ice machine in Jackson, Mississippi, in 1869.[71] After the war, Lowe was engaged in the high-volume production of hydrogen gas for balloons. He would amass a fortune on his other inventions, the water gas process and apparatus to produce hydrogen gas for balloons (and which later, was used in dirigibles), and for other products that ran on hydrogen gas. In 1886, he would be awarded the Franklin Institute's prestigious *Elliot Cresson Medal* for the "Invention Held to Be Most Useful to Mankind." In 1887, Lowe would move to Los Angeles, California, and open several ice-making plants.[72]

Chapter 3

Electricity and Invention

After the Civil War, the next challenge for early refrigeration machines, which were enormous in size and weight, was to adapt them to transport perishable goods over great distances, that is, by sea. It was Thaddeus Lowe who made one of the earliest attempts. Back in 1869, after Lowe had built his ice machine, using dry ice made with carbon dioxide compressors, he and others in Dallas purchased an old steamship, the *William Tabor*, into which they loaded one of his refrigeration units. They had entered a competition with Dr. Henry Peyton Howard of San Antonio to carry chilled and frozen beef to New Orleans. Howard's steamship, the *Agnes*, was fitted with a cold storage room, twenty-five by fifty feet in size. Because the *William Tabor* drew too much water to dock in New Orleans's harbor, Howard's was the first to ship refrigerated beef successfully by refrigerated boat. Upon arrival in June, Howard threw a banquet at the St. Charles Hotel in New Orleans, and presented his beef to prominent diners.[1] The New Orleans *Times Picayune* wrote poetically of the event: "[The apparatus] virtually annihilates space and laughs at the lapse of time; for the Boston merchant may have a fresh juicy beefsteak from the rich pastures of Texas for dinner, and for dessert feast on the delicate, luscious but perishable fruits of the Indies."[2] Because Lowe failed to complete his feat, he did not receive the proper credit for his attempt; however, the accomplishment of a refrigerator ship established the compressor process of refrigeration for ships delivering meat to New York and Europe. Carbon dioxide is non-toxic and non-inflammable, and it would be used in marine service well into the 20th century.[3]

James Harrison, the Australian who had developed a successful ice-making machine in 1854, and Thomas Mort, who had established the first meat-freezing works in Australia, both attempted the shipping of meat. They realized the enormous commercial possibilities of a potential frozen meat trade. In 1873 Harrison tried to adapt his equipment for an experimental

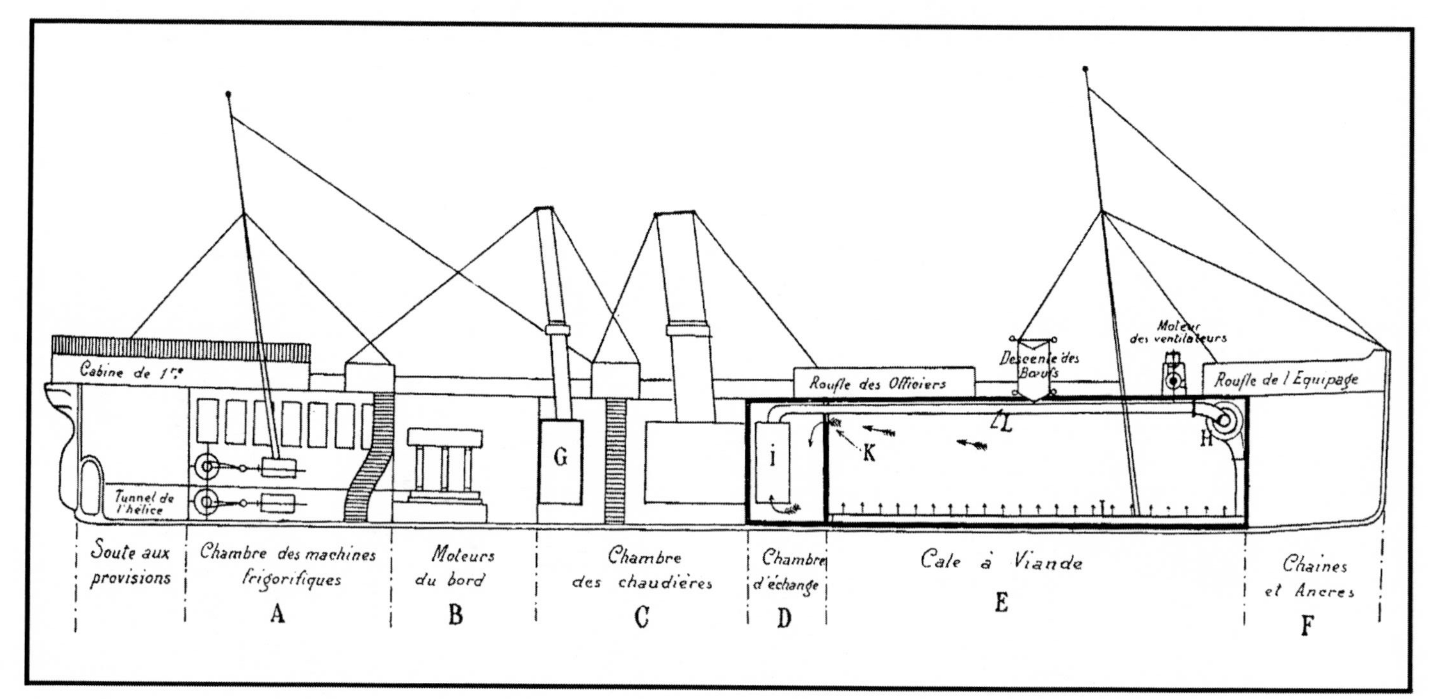
Cabine de 1re
Tunnel de l'hélice
Roufle des Officiers
Descente des Bœufs
Moteur des ventilateurs
Roufle de l'Equipage
G
i
K
L
H
Soute aux provisions
Chambre des machines frigorifiques
A
Moteurs du bord
B
Chambre des chaudières
C
Chambre d'échange
D
Cale à Viande
E
Chaines et Ancres
F

beef shipment to the United Kingdom in the sailing ship *Norfolk*.[4] Aboard were twenty tons of mutton and beef kept cold by a mixture of salt and ice.[5] Instead of installing a refrigeration machine aboard the ship, Harrison had relied on this cold storage system with ice, which proved disastrous when the ice melted faster than expected and the cargo was ruined.[6] Discouraged, Harrison left the freezing business forever.

Such failed shipping experiments with refrigerated meat lessened public confidence in this oceanic enterprise, but Australian, Argentine, French and American concerns continued to experiment with refrigerated shipping, and ice-making machines continued to be refined and developed. In 1875 Swiss inventor Raoul Pierre Pictet (1846–1929) developed the first machine using sulfur dioxide as a refrigerant. He was trying to address the problems of high-pressure ammonia machines when used in tropical climates. His machines would be sold in a number of industrialized countries by the 1880s.[7]

Finally, in 1877, frozen mutton was successfully sent in the ship *Paraguay* from the Argentine to Le Havre, France. Ferdinand Carré had equipped the *Paraguay* with ammonia-compression machinery of the type developed by Harrison in Australia.[8] Due to an unfortunate collision, the voyage took six months, but the hard-frozen meat still remained in excellent condition. Earlier, in 1875, Australian Thomas Mort had tried a different system for shipping, using ammonia as the coolant, and departed again with a shipment on the SS *Northam*, but a leak ruined this cargo as well. Eventually, Mort discovered the right system to survive the long voyage to London, when he went back to Gorrie's "dry air" system, as used on the Bell-Coleman machine, and the first meat to be chilled in this manner left Sydney, Australia, in December 1879 aboard the SS *Strathleven*, a nine-week voyage of about 15,000 miles, arriving on February 2, 1880, with the cargo of 40 tons of frozen beef and mutton intact. The meat sold at the Smithfield Market for about 5d per pound, and Queen Victoria (1819–1901), sampling a leg of lamb from the consignment, pronounced it excellent.[9]

Also in 1879, French inventor Charles Tellier refitted an English packet boat re-named *Le Frigorifique* with his cold storage refrigeration equipment to successfully carry meat and poultry chilled to 0°C from Argentina, South America, to France. His equipment used a shell-and-

Opposite: ***Le Frigorifique refrigerator ship, 1879*** **(Compte Rendu Officiel de la Manifestation Internationale** ***[Mâcon, France: Imprimé sur les presses de la Maison Protat frères, 1913]*****).**

tube heat exchanger to refrigerate air, blowing the air through the heat exchanger with a steam-driven fan. The process was described at the time as "refrigeration by cold, dry, air." But the return trip was not as successful, and Tellier did not continue to commercialize the idea.[10]

Still, the use of mechanical refrigeration to ship perishables by sea continued unabated. The ships outfitted with such equipment were called "reefer ships." A compressed-air refrigerator was fitted to a British ship, which in 1880 returned to London from Australia with 40 tons of frozen beef and mutton. That same year a sailing ship with the same equipment carried the first mutton and lamb from New Zealand.[11] This shipment was initiated in 1881 by William Soltau Davidson (1846–1924), a Canadian entrepreneur who had emigrated to New Zealand, and who by 1875 was superintendent of a sheep ranch in Timaru, New Zealand. At the time, sheep were raised only for wool, and wool prices had fallen by one third. Davidson thought that Britain's rising population and increasing demand for meat could solve the slump in world wool market that was affecting New Zealand.

New Zealand had attempted to export canned meat in the 1870s, but this didn't sell well in England. Live shipment of sheep was much too expensive. The first attempt to ship refrigerated sheep meat from Texas to New York by Thaddeus Lowe in 1869 had resulted in the loss of the entire cargo, and as we have seen, other attempts had varying success. After research into different methods, Davidson fitted the New Zealand clipper ship *Dunedin* with a steam-powered Bell-Coleman compression refrigeration unit, which used 3 tons of coal per day, and despite the loss of 650 carcasses due to a broken compressor, in 1882 the *Dunedin* sailed from Port Chalmers, New Zealand, for London with what was considered the first commercially successful refrigerated shipment. On board were 4,331 mutton carcasses, 598 lamb carcasses, 22 pig carcasses, and 246 kegs of butter, as well as hare, pheasant, turkey, chicken and 2,226 sheep tongues. It arrived in London after 98 days with its cargo still frozen, founding the industry and giving New Zealand an early lead, which led to a meat and dairy boom in Australasia and South America. Within 5 years, 172 shipments of frozen meat would be sent from New Zealand to the United Kingdom.

The refrigerated meat trade to Great Britain increased when J&E Hall of England in 1886 outfitted the SS *Selembria* with a vapor compression system to bring 30,000 carcasses of mutton from the Falkland Islands.[12] By 1899, refrigerated ship traffic to the United States reached 90,000 tons per year, and a 1900 worldwide survey found 356 refrigerated ships, 37 percent of which had air machines, 37 percent ammonia compressors, and 25 percent carbon dioxide (CO_2) compressors. The leading shipper was the United Fruit Company, which had been using "reefers" since about 1889, often combined with cruise-ship passenger accommodations to pay for part of their operating expenses. Because their cargo was mostly bananas, they have been called the "Banana Fleet," but they were in fact the start of what we now call cruise ships. After 1910, they would call their combination cruise and refrigerated cargo ships the "Great White Fleet." To avoid U.S. shipping regulations and taxes, the ships were registered in about six other countries, rather than in the United States. In the 1980s, United Brands would be taken over by Chiquita Brands International in Cincinnati, which owns the largest fleet of banana boats in the world. Frozen meat and dairy exports would continue to form the backbone of New Zealand's economy until the United Kingdom's entry into the European Economic Community (EEC) in 1974 would lead to New Zealand produce being excluded by the EEC's trade bans. Today, New Zealand has no refrigerated cargo ships.

In the 1870s, artificial ice-making technology began to be used for purposes other than the shipment of perishables, beer brewing, and cold storage. It was now being used for recreational purposes. The recreation was ice-skating, a popular pastime throughout the world for centuries during winters in temperate climates, as people flocked to frozen canals or lakes to enjoy the sport. Early attempts at artificial ice rinks were made in the early 1840s during the "rink mania" of the time in London, using a mixture of hog's lard and various salts, but people soon tired of the smelly ice substitute. But as refrigeration technology developed, John Gamgee of England conceived of a process while attempting to develop a method to freeze meat for import from Australia and New Zealand, and patented it in 1870. His process for an artificial ice rink consisted of a concrete surface, with layers of earth, cow hair, and timber planks. Atop these were laid oval copper pipes carrying a solution of glycerin with ether, nitrogen peroxide, and water. The pipes were covered by water and the solution was pumped through, freezing the water solid into ice.

Gamgee opened the first mechanically frozen ice rink, the Glaciarium, in a tent in a small building just off Kings Road in Chelsea, London, on January 7, 1876. In March, it moved to a permanent venue at 379 King's Road, where a larger rink measuring 40 by 24 feet was established. Gamgee operated the rink on a membership-only basis, attempting to attract a wealthy clientele experienced in open-air skating in the Alps. He installed an orchestra gallery, which could also be used by spectators, and decorated the walls with spectacular murals of the Swiss Alps. The rink proved to be a great success, and Gamgee opened two additional rinks later that year at Rusholme in Manchester and at the Floating Glaciarium in Charing Cross in London, the latter being much larger at 115 by 25 feet. Gamgee patented his method in 1877 (U.S. Patents 208,304 and 208,305). The Southport Glaciarium would open in 1879, using Gamgee's effective method.[13] In the 1890s, ice-skating rinks, made possible by mechanical refrigeration, were constructed in many major world cities.[14] Today's artificial ice rinks use very similar methods, with minor differences in refrigerants and with promotional logos and boundary markings embedded in the ice for sports such as ice hockey.

In what was by now the refrigeration industry, Alabama inventor David Boyle, after years of experimentation, had finally established his first ammonia-compression machine plant in a lean-to off a lumber mill in Jefferson, Texas, in 1873. Improvements made that winter of 1873–1874 resulted in a high-grade production that attracted national attention, but the machine was destroyed in a fire in 1874. He had finally received financial backing that year for his ice machines, which he patented in 1875. In 1877 he moved to Chicago, Illinois, where he contracted with Richard T. Crane of Crane and Company to manufacture his improved ammonia vapor-compression machines and founded the Boyle Ice Machine Company. His first two machines were bought by the Capitol Ice Company of Austin, Texas, and by Richard King (1824–1885), who wanted to experiment with meat refrigeration on his famous King Ranch.[15] Boyle sold a refrigerating plant to the Bemis and McAvoy Brewing Company in Chicago, and his systems would be manufactured by his and successor companies until 1905.[16]

Another early worker in the development of ice-making machinery was Charles A. Ziker of San Antonio and Austin. After he came to Austin from Indiana in 1880, he worked in an ice plant that had been using a Carré machine brought from San Antonio. Richard King in 1882 asked Ziker and his brother, Andrew J., to go to Brownsville and operate a Boyle ammonia-compression machine at an ice plant King had bought in 1876. Ziker returned to Austin in 1884, built his own plant, and continued to improve and design compressor-type ice-making machinery. In business with George W. Brackenridge, a San Antonio banker, Ziker established ice plants in Austin and San Antonio. After that he built plants in any city where he could find enough prosperous people and sufficient cooling water for compressors. In 1928, he would sell his ice plants, which ranged from Texas eastward to Atlanta and northward to Pittsburgh, to the Samuel Insull interests for $1 million.[17]

But meanwhile, in the U.S. natural ice industry in 1870, another ice famine had occurred, and then yet another in 1880. By that time, an estimated eight million tons were being harvested annually. Despite this, during the 1880s, the huge demand for ice continued to expand, and the cost of sawdust, shavings, and marsh hay skyrocketed. About four million tons of ice were routinely stored along the Hudson River banks in 135 major ice warehouses employing over 20,000 workers. However, as a result of these ice famines every decade, which created huge shortages and raised prices, ice entrepreneurs continued to expand operations on the Kennebec River in Maine as an alternative source to New England and the Hudson Valley. The Kennebec, as well as the Penobscot and Sheepscot rivers, were opened for the ice industry and became the most important source of ice for the rest of the century, as ice famines in New England hit again in 1890 and the Hudson River harvests failed completely.[18] The ice famines of the 1800s sound a bit like the global warming predictions of today, but they did nothing to discourage the growing demand for natural ice. These ice famines, along with increased cost of insulation materials, were more than balanced by increased demand, and only spurred the manufactured ice industry forward, increasing its competitive advantage, annual volume of production, and potential profits.

Manufactured ice-making technology changed dramatically in 1880 with the evolution of the electric motor as a new, lightweight, power source. That year, an Edison motor was used on a commercial sewing machine, and in 1882, Philip Diehl of New Jersey used such a motor to power a ceiling fan, which he sold as the Diehl Electric Fan. Electric motors have been used for ceiling fans ever since and are still popular in the South. Portable electric fans would soon become the first electric consumer products for the home, and in 1889, Singer would introduce the first practical electric sewing machine.[19]

The electric motor had been evolving since 1821, when English physicist Michael Faraday (1791–1867) discovered that electromagnetism could produce motion by electricity flowing through one circuit to induce current in another. He built the first experimental electric motor, essentially a rotating needle. In 1835, Brandon, Vermont blacksmith Thomas Davenport (1802–1851) built a model car powered by a small, battery-powered electric motor, and is credited for the invention of the first direct current (DC) electric motor, patented in 1837. In 1845, Charles Wheatstone developed the first electric generator, and in France, electric storage batteries were developed and used to power electric vehicles after 1865.

In 1884, Frank J. Sprague (1857–1934), a former employee of Thomas

Edison (1847–1931), founded the Sprague Electric Railway and Motor Company, and in 1886 invented a DC, constant speed, non-sparking motor for electric streetcars. This proved that electric current could be used for motors with stationary and rotating magnets, but DC motors required a moving commutator to reverse the direction of the current, and brushes to transmit the current to moving parts, both of which wore out quickly. But the greatest disadvantage of DC current was that it could not be transmitted over any great distance, because it lost electrons through friction as it traveled through the wires, and it could not be easily converted to higher or lower voltages. This would become a huge problem for Edison, who had had formed the Edison Electric Light Company in 1878, had invented the first commercially practical electric light bulb in 1879, and had, in 1882, switched on his Pearl Street Station in New York to provide 110-volt direct current to 59 customers in lower Manhattan. But with DC current, electricity could only be transmitted a short distance, severely limiting its potential for large grid systems.

Serbian inventor Nikola Tesla (1856–1943) solved this major problem for the distribution of electricity. Tesla had worked for the Continental Edison Company in France in 1882 and, upon his arrival in the United States in 1884, was hired by Edison. Tesla grandly claimed he could redesign Edison's inefficient motors and generators, making improvements in both service and economy. According to Tesla, Edison remarked, "There's fifty thousand dollars in it for you—if you can do it." After a year of working to redesign Edison's dynamos, Tesla went to Edison to announce success, and requested his $50,000. "Tesla," Edison replied, "you don't understand our American humor," implying that his offer was a joke. Instead, Edison offered Tesla a $10-a-week raise over his $18-per-week salary. Feeling cheated, Tesla refused the offer and immediately resigned.[20]

Tesla then established his own Tesla Electric Company in 1887 and proceeded to perfect a completely different alternating current (AC) system that he had originally conceived in Budapest in 1882. In the AC system, the movement of the electric current itself periodically reverses direction (60 times per second), and the voltage can be easily increased or decreased with a transformer, so the same amount of power can be transmitted with a lower current by increasing the voltage (by hundreds of kilowatts). This leads to more efficient transmission of power through wires, enabling much greater distances for electricity distribution.

Perhaps even more importantly, in 1889, Tesla invented an AC motor that consisted of only two basic parts: an outside stationary stator to produce a rotating magnetic field, and an inside rotor attached to the output shaft that was given torque by the rotating field. The rotating field was created by two or more alternating currents out of step, or phase, with one another, and was thus called a polyphase system. His motor was called an induction motor because an induced current created the magnetic field in the rotor. The advantage over the DC motor was that it eliminated the need for the brushes and commutator, providing a much longer motor life and making AC motors much more feasible for industrial use. Tesla was granted 40 patents on his polyphase system between 1889 and 1891, and leased his patents for royalties to George Westinghouse (1846–1914) to use on trains and streetcars. However, because of controversy regarding the safety of alternating current, DC current and motors would still be produced for the next 20 years, and both systems would compete with each other, the proponents of each hoping that their preferred system would prevail.

The controversy was the result of what became called the "War of the Currents" between Thomas Edison (promoter of DC, direct current) and George Westinghouse (promoter of AC, alternating current). Edison did not want to lose the royalties he was earning from his direct current patents, and began a campaign to discredit AC. He spread information saying that AC was more dangerous, even going so far as to publicly electrocute stray animals using AC to prove his point. He also tried to popularize the term for convicts being electrocuted for a death sentence as being "Westinghoused." Although Edison opposed capital punishment, his desire to disparage the use of AC led to the invention of the electric chair. He secretly paid Harold P. Brown to build the first electric (AC) chair for the state of New York to promote the idea that A.C. was deadlier than DC. When the chair was first used on August 6, 1890, the technicians on hand misjudged the voltage needed to kill the condemned prisoner, William Kemmler. The first jolt of electricity was not enough to kill him, and only left him badly injured. The procedure had to be repeated and a reporter described it as "an awful spectacle, far worse than hanging." George Westinghouse commented, "They would have done better with an axe."[21]

Meanwhile, experiments with long-distance transmission of electricity were ongoing. In 1889, the first, with DC, was switched on at the Willamette Falls station in Oregon City, Oregon. Unfortunately, a flood destroyed the power station in 1890, paving the way for the first long-distance transmission of AC electricity in the world when Willamette Falls Electric Company installed experimental AC generators from Westinghouse. That same year, the Niagara Falls Power Company (NFPC) formed the International Niagara Commission of experts to analyze proposals to harness Niagara Falls to generate electricity. In 1893, the NFPC awarded the contract to Westinghouse and rejected the Edison General Electric Company's, and Edison's, proposal. Power was to be generated and transmitted as AC.

At the Chicago World's Fair in 1893 (also known as the Columbian Exposition, the "White City," or the first Electrical Fair), General Electric, created in 1892 from Edison's and other electric companies using DC current, competed with Westinghouse to provide power and equipment for the fair, and Westinghouse underbid to obtain the contract for $399,000 (General Electric bid $554,000). The fair used Tesla's alternating current system, and Tesla himself presided in the electrical building, demonstrating one electrical miracle after another for entranced crowds. In 1896, based on Tesla's inventions, Westinghouse completed the hydraulic power station at Niagara Falls, encouraging industrial use of electric power. Westinghouse had essentially won the War of the Currents. By 1897, Westinghouse was so successful that royalties accrued to Tesla for his inventions reached about $12 million, and would have grown to billions, ruining Westinghouse if continued. George Westinghouse offered to buy Tesla's patents for $216,000 to avoid paying for royalties, and out of gratitude to Westinghouse for developing his inventions, Tesla generously agreed. Ironically, Tesla would die in relative poverty.

Meanwhile, research and development of refrigeration machines had continued in Germany in 1870 by Carl Paul Gottfried Linde (1842–1934), professor of mechanical engineering at the Munich, Germany, Technische Hochschule (Technical College). Linde published a paper describing his findings: "The Extraction of Heat at Low Temperature by Mechanical Means."[22] In 1871, he published another paper: "Improved Ice and Refrigerating Machines."[23] Linde had applied for a job as lecturer at the school in 1868 and became a full professor of mechanical engineering in 1872. He set up an engineering laboratory where students such as Rudolf Diesel (1858–1913, inventor of the diesel engine) studied. Linde's first refrigeration units were vapor-compression machines, built by Maschinenfabrik Augsburg for the Spaten brewery in 1873. The machines used dimethyl ether as the refrigerant and were commercially successful; as a matter of fact, they were twice as efficient as other machines of the time. In 1877 Linde switched to more reliable ammonia-based systems (U.S. Patent 228,364), which are still in use in industrial applications, and his improved model was installed at the Dreherschen Brauerei at Trieste, Austria. Beer was obviously still a major client for the mechanical refrigeration industry.

Carl Linde's first ammonia refrigeration system, 1877 (courtesy the Division of Work and Industry, National Museum of American History, Smithsonian Institution).

This new system incorporated a lower-cost compressor by combining the two single-acting cylinders into a double-acting one, and a more efficient sealing system. By 1879 Linde gave up his professorship in Munich and founded Linde's Ice Machine Company in Wiesbaden, Germany. In 1881, Chicago engineer Fred W. Wolf secured rights to manufacture and sell Linde's refrigerating systems in the U.S. (U.S. Patent Reissue No. 10,522). Linde's efficient technology offered large benefits to breweries, and by 1890, he had sold 747 machines in the American market,[24] and about 12,000 machines in Germany.[25]

In 1892, an order from the Guinness brewery in Dublin for a carbon dioxide liquefaction plant drove Linde into exploring low-temperature refrigeration; in 1895 he achieved the liquefaction (the state of being liquefied) of air, and he filed patents in Germany and the United States by 1903. His apparatus combined the cooling effect achieved by allowing a compressed gas to expand with a counter-current heat exchange technique that used the cold air produced by expansion to chill ambient air entering the apparatus. Over a period of time, this gradually cooled the apparatus and air within it to the point of liquefaction. In 1897, Linde was ennobled as Ritter (Knight) von Linde for his innovative work in refrigeration and in 1907 he formed the Linde Air Products Company in the United States, which eventually passed through U.S. government control in the wartime 1940s to the Union Carbide Company.[26]

It was primarily the beer-brewing industry that had added enough potential business to the ice-making industry and had provided a sound financial basis for the refrigeration business to continue to exist and flourish. From Harrison's 1861 machines in an Australian brewery to Linde's 1879 work with German and Irish breweries, breweries provided much of the business for refrigeration machines up to 1890. By 1895, an advertisement for the De La Vergne Refrigeration Machine Company listed 583 refrigeration plants installed, and 369 of them were breweries (over 60 percent).[27]

As we have seen, as the mechanical ice-making and refrigeration machines were being developed mechanically, a number of different refrigerants of high volatility were experimented with, the earliest being ether. Most of them saw only limited use. But those that did see long-term use were ammonia, carbon dioxide, sulfur dioxide, methyl chloride, and ethyl chloride. The history of the development and use of these can be described in more detail.

Ammonia is a compound of nitrogen and hydrogen with the formula NH_3. It is a colorless gas with a characteristic pungent smell, half the density of air, and boils at –33.34° C., so it must be stored under high pressure or low temperatures. Ammonia, as used commercially, is often called anhydrous ammonia, meaning there is no water in it; it is dry, as compared to household ammonia, ammonium hydroxide, which is a solution of NH_3 in water.[28]

In his patents, Alexander Twining specified ammonia as early as 1850, and James Harrison did so in 1856, but both used ethyl ether in the machines they actually constructed. Charles Tellier of France experimented with ammonia in 1862, but chose to use methyl ether for his early systems. Eugene Nicolle, the French engineer from Sydney, Australia, and a pioneer in the export of refrigerated meat, claimed that he made an ammonia compressor in 1863–1864 but did not develop it further because the absorption-type system looked more efficient to him. A British patent (3,062) was issued to R.A. Brookman in 1864 for an ammonia compression machine for ice-making and refrigeration. John Beath of San Francisco built several ice-making plants using ammonia compression systems in 1868–1872, but by 1873, he abandoned attempts with vapor compression systems, claiming they had too many problems, and turned to absorption systems.

The first to use ammonia successfully was probably the American Francis V. DeCoppet. George Mertz in New Orleans had installed a Tellier machine in 1869 that had serious problems with leaks (Tellier's machines were made for methyl ether, not ammonia). Mertz hired DeCoppet to correct the problem and he did so by constructing a double-acting ammonia compressor in 1870 (U.S. Patent 156,056, *Improvement in Ice Machines*). The compressor operated successfully for two years. John Enright of the United States constructed an ammonia compression system in 1876 (U.S. Patent 202,641, *Improvement in Ammoniacal-Gas Pumps*) that was installed at Ziegel's brewery in Buffalo, New York, in about 1877. Enright was an engineer for F.M. McMillan and Company, its successor, Arctic Machine Company of Cleveland, Ohio, and its successor, Arctic Ice Machine Company of Canton, Ohio, one of the earliest manufacturers of refrigerating equipment in the United States, which continued to manufacture ammonia systems into the 20th century. McMillan was the first to manufacture anhydrous ammonia commercially, beginning in 1876. The lack of availability of dry ammonia caused some machinery manufacturers to supply stills and lime dryers with their equipment so that anhydrous ammonia could be produced from aqua ammonia, which was commercially available. By the end of the century, a number of manufacturers would be producing refrigerant-grade ammonia.

Carbon dioxide (CO_2) (also known as carbonic acid gas or carbonic anhydride) was first proposed as a refrigerant by Alexander Twining, as mentioned in his 1850 British patent. Scientist and inventor Thaddeus S.C. Lowe had realized the possibility of carbon dioxide as a refrigerant in the 1860s, and successfully used it in his ice machines built in 1867. Carl Linde also experimented with carbon dioxide when he designed a machine for F. Krupp at Essen, Germany in 1882. W. Raydt received British Patent 15,475 in 1884 for a compression ice-making system using carbon dioxide. British Patent 1,890 was granted to James Harrison in 1884 for a device for manufacturing carbon dioxide for refrigeration purposes. But the use of carbon dioxide did not advance until Franz Windhausen of

Germany designed a carbon dioxide compressor in his British Patent 2,864 in 1886. Windhausen's patent was bought by J. and E. Hall of Great Britain, who improved it and started manufacture in about 1890. Hall's machines saw widespread use on ships, replacing the compressed-air machines previously used. In the United States, carbon dioxide was used successfully in the 1890s for refrigeration, and in the 1900s for comfort cooling. The principal user in the United States was the Kroeschell Brothers Ice Machine Company, which manufactured systems under patents purchased from the Hungarian Julius Sedlacek. Carbon dioxide systems were widely used on British ships in the 1940s, but after that they began to be displaced with chlorofluorocarbon (CFC) refrigerants.[29]

Sulfur dioxide (also known as sulfurous acid, sulfurous anhydride, or anhydrous sulfurous oxide) is a toxic gas with a pungent, irritating, and rotten smell, and its formula is SO_2. A.H. Tait in his U.S. Patent 94,450 proposed it as a vapor-compression refrigerant as early as 1869. It had the advantage of low cost and operating pressures, but pressure was high enough to prevent air entry into the system. Daniel L. Holden claims he experimented with sulfur dioxide refrigeration systems in about 1870 (U.S. Patent 101,876, *Improvement in Apparatus for the Manufacture of Ice*). But the earliest commercial use was by the Swiss Raoul Pierre Pictet, who perfected his machines in 1876. He had obtained British Patent 2,727 in 1875 and manufactured his machines through the Société Genevoise beginning in 1876 and after that through a successor firm in Paris. A unique feature of his machine was its lack of oil or grease. A 1911 article by W.S. Douglas in the October issue of *Ice and Refrigeration* remarked: "It follows that if the cylinder is water jacketed to promote slight condensation on its interior walls, no oil whatsoever need be fed to the piston."[30]

Sulfur dioxide was used more in Europe than the United States, usually in small to medium sized applications. It became increasingly popular after 1900, by those perfecting the electric household refrigerator.

Methyl chloride, a colorless, flammable gas with a mildly sweet odor, which can be toxic, is a gas synthesized from methanol, sulfuric acid, and sodium chloride with the formula CH_3Cl. It would later be banned from consumer products because of damage to the atmosphere. Methyl chloride was discovered in 1835 by French chemists Jean Baptiste Dumas and Eugène-Melchoir Péligot and used as a topical anesthetic for French battlefield amputations. Camille Vincent of France, who received British Patent 470 in 1879 using a two-stage compressor, used it as a vapor-compression refrigerant in about 1878. At about the same time, Cassius C. Palmer in the United States specified methyl chloride in his British

Pictet's first sulfur dioxide ice-making machine, 1875 (*Les Machines à Glace *by Auguste Perret, Paris, 1904, Figure 94).

Patent 1,752 of 1880, although he chose to use higher-boiling-point ethyl chloride in his final machines. Methyl chloride was used very little until World War I, when it was widely used in small electric refrigerators.[31]

Ethyl chloride, like methyl chloride, was also used as a topical anesthetic. It was used as a vapor-compression refrigerant starting about 1870. It is a colorless, flammable gas with a sweet odor with the chemical formula C_2H_5Cl. As mentioned above, Cassius Palmer used it in his final machines designed for refrigerating systems in railway cars in 1883, and he was the first to use a rotary-vane compressor for refrigeration purposes. By 1898 Palmer was operating his business out of Chicago as the Railway and Stationary Refrigeration Company of New York City. His systems used ethyl chloride under its trade name Clothel.[32]

The development of refrigeration was not just a matter of inventors, engineers, machines, refrigerants, processes, manufacture and distribution. Refrigeration technology created what refrigeration engineers call "cold chains." Cold chains consist of the linked refrigeration technologies, industries, and actions required to preserve and transport perishable food from its point of production to its point of consumption. It is similar to a food chain, with plants at the bottom and predators at the top. The same refrigeration equipment could be adapted to make ice, provide storage warehouses, and cool railroad cars or ships. All along the way, various industries were developed to take ice and foodstuffs on long journeys around the world and the ubiquitous household icebox refrigerator was the endpoint of this vast chain.[33]

New uses for refrigeration popped up everywhere in the 1880s. Undertakers saw an opportunity. At the time, embalming bodies of the deceased was not common. To prevent decomposition, various methods were devised to preserve bodies. At first, ice was used, but as mechanical refrigeration became available, inventors used it for "corpse cooling." In the 1880s numerous patents were issued for "corpse coolers." One of these was by African American inventor Thomas Elkins of Albany, New York, who was granted U.S. Patent 221,222). Towards the end of the 19th century, refrigeration by pipeline, or "district refrigeration," became fashionable. Central stations were connected to houses, stores, offices, etc., by underground pipe mains carrying chilled brine or ammonia. Such systems appeared in many cities, including New York, London, Paris, Boston, Louisville and Nashville. The first successful pipeline refrigerating system was installed in Denver, Colorado, in 1889. Another system was installed in St. Louis in 1891.[34]

While the mechanical refrigeration industry was progressing by leaps and bounds, this was not the case with natural ice. The latest "ice famine" of 1890 was a severe blow to the natural ice industry. The editors of a new trade journal, *Ice and Refrigeration*, first published in 1891, reported it:

> Nearly all will agree that the ice season of 1890 has had the most unfortunate results. The previous winter was an open one, and very little ice was harvested in the North. As a consequence, the supply in that section was not equal to the demand and a rise in price followed. This and the erection of ice machines in several of the larger cities of the North, furnished ample food for discussion by the newspapers of the country, and, as is usually the case ... a large amount of misinformation was scattered abroad.[35]

In addition, after 1890, natural ice was also encountering serious environmental problems. Because of the rapid growth of cities along rivers and lakes, the waters were increasingly polluted with sewage and refuse. Ice harvested from those waters was unsightly, smelly, and dangerous to health with concentrations of harmful bacteria. An article in the *Detroit Tribune* diligently warned:

> For a long time intelligent people have realized that sewage infected ice is rather more than it is cracked up to be. It is probable that many of the

> summer ailments of the digestive tract are due to the uses of villainously impure ice, which is loaded with dormant bacteria.... Some of them produce typhoid fever and other filth diseases that frequently end in death.... Frozen river water that is contaminated with sewage does not warn the consumer unless the filth is present in large chunks that will not strain through the teeth.[36]

In many areas, cities banned the sale of ice harvested from such sources. With the increased availability of plant ice, the sale of natural ice continued to decline. The natural ice industry fought back against these difficulties by trying to convince the public through a media battle that natural ice was better than manufactured ice, and by referring pejoratively to manufactured ice as "artificial ice," but others decided to join rather than fight, building ice plants to supplement or replace their natural sources.

Manufactured ice became the beneficiary of these problems, and gradually cut deeply into the once-predominant natural ice market. Mechanical production technology had progressed so much by 1890 that it became increasingly cost-effective.[37] By 1900, there were 766 ice plants in the United States, and 77 were in Texas, more than any other state. Most still used the aqua-ammonia cooling method.[38] Although such technology was still too unreliable, large, and costly to be used in smaller stores and homes, it was widely used in almost all breweries and ice factories.[39] All commercial ice-making machines were incredibly large and heavy. Few machines weighed less than five tons, while many were from 100 to 200 tons.[40] It was not yet even imagined that they could be reduced in size and cost for use in domestic homes. Traditional wooden icebox refrigerators, using natural or manufactured ice, were the only option available for domestic use.

Widespread use of household icebox refrigerators occurred first in the United States, based on the number of patents issued and the number of advertisements in the media. It seems to have begun in the 1870s, and usage was much higher here than in Europe. In 1892, an Italian province of one million people used less ice than a U.S. town of ten thousand. In England, as late as 1904, few household iceboxes were evident despite their use being strongly advocated by health authorities. By that time there was a huge array of icebox refrigerators available in the United States. The styles ranged from the simplest, unadorned wooden box to intricately carved and mirrored pieces of gaudy furniture made of fine woods. Some were poorly made, while others were of excellent engineering and construction.[41] Many manufacturers were building such iceboxes to store the ice in the home, usually the kitchen. The Belding-Hall Company in Belding,

Michigan, founded in 1877 by Joshua Hall, was one of these. The area around Belding had a skilled workforce of Danish craftsmen and a good supply of hardwoods, including ash. In 1900, Frank Gibson, a competing manufacturer of "ice refrigerators," bought the company and continued to manufacture iceboxes under the name of the Gibson Company. It was the largest in the industry at the time.[42] Numerous inventors submitted patents to improve iceboxes, including in 1891, African American John Standard, of Newark, New Jersey, who patented an improved refrigerator design, U.S. Patent 455,891.

Icebox refrigerators soon became one of the standard appliances in the average kitchen, along with the stove and the sink; all essential to the preservation, preparation and cooking of food. A well-equipped kitchen was then, as today, a mark of a family's social position and standard of living. But the icebox was not carefree. It required considerable physical effort in supply and maintenance to insure effective operation. Ice dealers sold customers ice coupons, which could be used in lieu of cash upon delivery, and dealers also provided customers with free ice picks to chip away small pieces from the large block to place in pitchers of water or other beverages. Customers were also provided window cards with large printed numerals (10, 15, 20, and 25, etc.). By placing the card on the inside window ledge and rotating the card so that the desired weight of ice was at the top, the customer could automatically inform the ice man, who could see the card from the street, and know what specific weight ice block to deliver to that particular house.[43] Ice men were burly, strong fellows, as ice cakes for large iceboxes could weigh up to 300 to 400 pounds.[44]

In more rural areas without door-to-door delivery, ice was delivered to local "jitney stations," small insulated buildings owned by ice dealers, where ice was dispensed on the spot to customers (often children) who loaded the ice into whatever means of transportation available, usually small wheeled carts or wagons called "jitneys," for transport to individual homes. After the 1880s, the use of iceboxes in private homes increased dramatically, as the population increased and living standards rose.[45] Although some models had spigots for draining melted ice water from a catch pan or holding tank into a drain, most had a drip pan placed under the icebox that had to be removed and emptied daily. The user had to replenish the ice, daily or perhaps twice weekly, by purchasing ice from the local iceman, who became as common a social institution as the mailman, the milkman, the butter and egg man, and the coal man. Typically, the ice was cut into blocks of 25, 50, and 100 pounds, sometimes less, and

Typical window card to order ice (painting by, and courtesy of, Joseph J. Schott, Fanwood, New Jersey).

delivered in horse-drawn wagons, from which the iceman would carry the block of ice with large tongs to the customer's icebox. Some apartment buildings had small doors that opened directly into the icebox from the back porch. The iceman would bring the block of ice up to the porch and place it into the icebox through this door.[46]

The block of ice was effective in keeping the contents of the icebox cool, as intended, but ice customers had other needs for ice in smaller quantities, for example, to add small ice chips to beverages or freshening lettuce for salads. For this purpose, the ice supplier provided users with free ice picks: hand-held, dagger-like tools with a sharp point (sometimes used as lethal weapons; permits were not required). Customers could use them to chip off a small chunk of ice, place the chunk in a sink or dishpan, and use the ice pick to break it into smaller chunks to add to their beverages or for other household purposes.

Another use for these smaller ice chips was to treat bruises, black eyes, or hangovers, using rubber ice bags filled with chipped ice. Most homes had such implements in the bath closet alongside the rubber hot-water bottle, which was filled with hot water and used to apply heat to sore muscles or aches. They were both made of rubber, but the ice bag, in order to accommodate chunks of ice, had a screw-top opening about three inches in diameter. One filled the bag with chunks of ice and screwed the lid onto the threaded collar opening, so that melting ice water could not leak out. The bag of ice was placed on the injured or sore area until the ice melted and the bag needed refilling. Later, when household refrigerators became available, similar ice bags were filled with ice cubes from the mechanical refrigerator. Such ice bags are still available today, used to treat bruises or soreness.

Iceboxes, of course, required insulation around the ice to delay the melting of the ice within and the entry of warmer air from without. As we have seen historically, various materials such as tree branches, cloth, fur, sawdust, shavings, and marsh hay had been used over the years to serve as insulation. But all of these are less than ideal insulators. It was in 1904 that a device appeared on the market that provided maximum insulation by using absolutely "nothing." It was called a Thermos, and although quite simple in design and operation, and available at very little cost, it provided the best means possible of keeping things, usually liquids, cool or warm for many hours. It came about because of James Dewar (1842–1923), a Scottish chemist and physicist who conducted extensive research into the liquefaction of gases, including oxygen and hydrogen, at temperatures approaching absolute zero. To keep these liquefied gases at low temperatures for longer periods of study, in 1892 Dewar designed a vacuum-jacketed container (a

*Iceman delivering ice (*Ice and Refrigeration*, July 1908, 23).*

double-walled flask with a vacuum between two silvered layers of steel or glass). Dewar never patented his invention, which allowed a newly formed German company, Thermos GmbH, to take over the concept and develop a commercial version of the vacuum flask, which was soon a popular consumer item called a Thermos bottle, handy for keeping ice tea cool, or hot tea warm, for long periods of time, without any refrigeration or moving parts whatsoever.[47] Our turn-of-the-century ancestors thought it was better than—well, warm ice tea or cold coffee. You might ask: if a vacuum is such efficient insulation, why could it not be used to insulate a refrigerator? As a matter of fact, it is. Today, vacuum insulation is used in some Japanese refrigerators, but the insulation material is not yet durable enough to last the life of a refrigerator, because air molecules can diffuse through the plastic envelopes. To be as effective as a thermos bottle, it would have to be contained in a sealed glass container that was silvered. Such a large, odd-shaped container would be prohibitively expensive to manufacture.

The arrival of the 20th century called for a national celebration. The 1904 international Louisiana Purchase Exposition in St. Louis, Missouri, a world's fair, fit the bill. The Louisiana Purchase was actually made in 1803 but the fair was delayed past the anniversary to allow for full participation by more states and foreign countries. Along with promoting many other technological wonders of the new century, the fair encouraged the usage of ice and iceboxes by homeowners. Mechanical refrigeration, as well as electricity and electric motors, were enthusiastically promoted as the latest technologies to serve mankind. A dramatically presented, glass-enclosed exhibit of a refrigerated room featured a life-size cow sculpted in butter, which lasted from the opening on April 30 to the fair's closing on December 1. Among the over 19 million fairgoers were many who were already using natural ice, but there must have also been others without iceboxes who could not avoid being reminded that their own butter softened and spoiled within days.[48]

Over the final quarter of the 19th century, the heroes of industrialization were the inventors, who created the basic technologies and practical applications that brought to the grateful public a plethora of mechanical and electrical devices that elevated the standard of living; provided new communication options, new transportation means, and entertainment; and started dozens of new, enormous, job-producing industries. From 1876 to 1900, the number of U.S. patents increased by 50 percent. Even a short list of the inventors is incredible:

Thomas Edison (1847–1931) invented the quadruplex telegraph (1874), the phonograph (1877), the carbon microphone (1878), the first practical electric light bulb (1897), the electric distribution system (1880), the kinetoscope (a motion picture camera, 1891), and the two-way telegraph (1892); Alexander Graham Bell (1847–1922) invented the telephone (1876); Nikolaus Otto (1832–1891) invented the first practical gasoline internal combustion engine (1876); Nikola Tesla (1856–1943) invented alternating current (1888), the electric induction motor (1889), and wireless telegraphy (1893); George Eastman (1854–1932) invented roll film and the Kodak camera (1888); Frank Sprague (1857–1934) invented electric streetcars (1880) and electric elevators (1892); Hiram Maxim (1840–1916) invented the machine gun (1884); and Reginald Fessenden (1866–1932) invented wireless transmission (practical radio, 1900). These are but a few of the best-known inventors and inventions of this era. The automobile and the Wright brothers' airplane of 1903 were just the icing on the 19th century cake. But invention alone was only the starting point in the refinement of industrialization, mass production, and 20th century modernity.

It is virtually impossible for us in the 21st century to appreciate the impact these inventors had on American society and education, but we can get a rough sense of their prestige and public admiration by multiplying, many times over, the impact that Bill Gates and Steve Jobs, inventors of our own era, had on the computer world. Gates and Jobs brought the age of computers into our personal lives and homes, but the 19th century inventors had much more of a dramatic impact on the daily lives of our ancestors. They transformed an agricultural age into an industrial age, and transformed the United States from a backwater English colony into a major world power with the world's highest standard of living. While the personal computer spawned a massive industry and thousands of derivative products, 19th century inventors created mechanical and electrical industries, not just in one field but in dozens, and along with them, the derivative mechanical and electrical products that went with them.

And just as computers in the late 20th century inspired an entire generation of young, technically oriented, computer nerds, the inventors inspired an entire generation of engineers in the 19th century to design and manufacture new products on a previously unimaginable scale of mass production. Engineers not only had to design the products and continue to refine and improve them, but also had to select materials in huge quantities and design the incredible machinery needed to produce them by the millions. Engineering became the wave of the future, the mechanical equivalent of the space age, and engineers were the ones who led it. In 1870 there were only 17 schools in the United States teaching engineering, but by 1890 there were 110, and by 1900, there were already 45,000 trained engineers, ready to advance the electrical and machine ages into the 20th century. From 1870 to 1914 the annual number of engineering graduates leaped from 100 to 4,300.

The refrigeration industry was only a small segment of this vast evolution of industrialization in the late 19th century, but as we have seen, there were a number of inventors and engineers who brought mechanical refrigeration into being, and many sources cite Oliver Evans as the inventor who first described in 1805 how such a system could work. Earlier, this book described the improvements of refrigeration by John Hague, Dr. John Gorrie, James Harrison, Alexander Twining, Ferdinand Carré, Carl Linde, and Charles Tellier, among others. The heating and ventilating industry first organized in 1894 as the American Society of Heating and Ventilating Engineers (ASHVE). In 1904, specialized engineers, many of whom had refined refrigeration into practicality and usefulness, organized the American Society of Refrigerating Engineers (ASRE). Many of them were also manufacturers, who, the year before, had formed the Ice Machine Builders Association (IMBA).[49] Their work over the next 50 years would make refrigerators and air conditioning a normal part of almost every household. Miniaturization and cost reduction, just as in computers, would be the key to success both in the commercial and in the domestic arena. In the next chapter, the development of various systems of comfort cooling will be described as well as a number of early electric domestic refrigerators, which would be the smallest ice-making machines ever.

Chapter 4

Early Electric Refrigerators

Although electricity and electric motors had been around since 1837, it took considerable time and energy to build the extensive distribution systems and transformers necessary for electricity to reach factories and homes, and to improve motors to be powerful enough to challenge the efficiency and cost of steam or gasoline engines, both primary sources of energy in the 19th century. It was not until about 1900 that electric motors began to effectively replace steam power in industry. Steam engines were still the predominant power source for huge refrigeration machines, of course, but they, just like the machines they powered, required the full-time care of a mechanic, large amounts of floor space, and large financial investments. Gasoline engines had also been around since 1876, but they were unsuitable for any indoor operation because of the toxic fumes. In contrast, electric motors required no mechanic, consumed less floor space, emitted no fumes, and were much less expensive. Some ice machines used electric motors in 1895 as they became available, but as late as 1902, although 1,447 refrigeration machines were in operation, only 85 of them were operated by electric motors. Steam engines would still be used in some refrigeration machines as late as 1920, when more reliable and more powerful electric motors would become more commonplace.[1]

It also took considerable time for air conditioning, long a passionate desire for hospitals and public buildings, to become a reality. John Gorrie had conceived the idea of "an engine for ventilation, and cooling air in tropical climates" in 1844, but never actually realized his dream. Although many in the 19th century devised systems that cooled the air through ventilation, evaporation, and ice, true air conditioning was not possible until mechanical refrigeration was successfully developed. As late as July 1881, when President James A. Garfield (1831–1881) was bedridden in the White House, suffering from an assassin's bullet, the best technology available for air conditioning to moderate the oppressive heat of the summer was a machine built by naval engineers in which an air blower was installed over a chest containing six tons of ice, with the air then dried by conduction through a long, iron box filled with cotton screens, and connected to the room's heating vent. The Rube Goldberg device could lower the room temperature by 20 degrees below the July heat outside, but used half a million pounds of ice in just two months. Garfield died anyway of a heart attack, blood poisoning, and pneumonia in September. Some generously regard this Garfield machine as possibly the first room air conditioner, but it was quite similar to the technique Dr. Gorrie had used in hospitals in the 1830s, 50 years earlier and 20 years before Gorrie patented his own ice machine.

Early advanced cooling systems in public theaters in the 1880s were not that different in principle from that used for Garfield; they only differed in scale. They not only used an enormous amount of natural ice to provide air conditioning, but also required complex and costly construction to handle the ice and air. As an example, in 1880, New York's Madison Square Theater was using four tons of ice to cool patrons at each individual summer performance. Fresh air was filtered through a 40-foot-long cheesecloth bag, passing over wooden inclined racks containing two tons of ice, and into an 8-foot-diameter centrifugal fan. The fan discharge was directed over another two tons of ice into ductwork to various openings through which the cool air "poured into the house to reduce the temperature and to furnish a supply for respiration." This complex system was engineered by the B.F. Sturtevant Company, an example of the growing influence of heating and ventilating engineers in the comfort cooling industry.[2]

In the 1890s, the standard commercial heating system that lent itself to easy use for room cooling was the "hot blast" system, which used ventilating fans to blow air over heated surfaces. A system using a combination fan and heat exchanger had been developed and patented (U.S. Patent 92,460) in 1869 by American Benjamin F. Sturtevant, who shortly thereafter established a company to manufacture and sell fans and heating systems. Such a system lent itself to cooling because the heating surface could be cooled in summer with refrigerated brine. By the 1890s, these 'hot blast' heating systems were being equipped with air washers for filtering and humidifying, and the air washer could easily be adapted for cooling purposes, although there was little domestic demand for comfort cooling in the 1890s because of the cost and size of the equipment required. Optimistically foreseeing the potential for effective air-conditioning, the editors of the *Stationary Engineer* in 1891 wrote: "The time is not far distant when mechanical refrigeration machinery will be applied to the cooling of hotels and dwellings in the summer in a manner similar to that by which the heating is now done in winter, and a portion of the heating apparatus will be used for this purpose."[3]

There was, indeed, increasing demand for cooling by commercial and public establishments, and many different systems were tried. In 1891, the St. Louis Automatic Refrigerating Company installed a district refrigeration system by which ammonia was circulated from a central plant through street pipelines. One of its customers was a restaurant and beer hall called the Ice Palace, which was cooled in summer by frosted pipe coils on the restaurant walls. The walls were decorated with painted murals depicting sleighing scenes, and "other frescos of a frigid character." One of these frescos was of the Kane polar expedition. Elisha Kane, 1820–1857, was assistant surgeon in the navy and a member of the 1855 Arctic expedition on the icebound brig *Advance*, which was rescued by a ship after an 83-day march over ice to safety.

Trying another method, the Kansas City Cable Railway Company experimented with cooling by using compressed air to cool passenger cars in 1891. Compressed air, originally created by manually operated bellows in mines to provide fresh air, became the basic principle of John Gorrie's refrigeration machine of 1845; compressed air could provide cooling when it expanded. By 1850, his machine was making ice, but compressed air systems were soon replaced by the more efficient evaporation of liquids. Mechanical air compressors took a different direction in the 1860s and 1870s, to create power rather than cooling. They were used to create a power source for mechanical tools that drove pistons that operated powerful rock drills, speeding up tunnel construction in the Alps. In the late 19th century, major cities such as Paris, Birmingham, and Buenos Aires had compressed air systems in place, the original purpose of which was to power clocks by providing a pulse of air every minute to advance the minute hand; but the systems also evolved into delivering power to homes and industry. At the time, many people regarded compressed air as a cleaner energy and power source than electricity. Indeed, compressed air would become a major source of power in many industrial applications, such as power tools, spray painting, and cleaning equipment.

The most intense commercial demand for cooling came from large public buildings, which were the most challenging installations because of the large number of people in vast interior spaces. In 1892, heating and ventilation engineer Alfred R. Wolff of Hoboken, New Jersey, installed an ice-type cooling system at the New York Music Hall (now Carnegie Hall). The building's dedication records described the complex system:

> Fresh air, at any temperature desired, in large volume but low velocity, is introduced ... and exhausted. Generally, fresh air (warmed or cooled) enters through perforations in or near the ceilings, and the exhaust is effected through registers or perforated risers in or near the floors.... Through the heating surface, or at will through the ice racks, the air is drawn by four powerful blowers [powered by steam], each 12 feet high, and forced through the system of fresh air ducts into various parts of the building. The heating surface and other appliances are so subdivided that atmospheric changes can be immediately compensated for, and the temperature of the air introduced suited to the winter weather or the heat of summer.[4]

A number of similar ice-type cooling systems were installed in public spaces during the early 1890s and early 1900s, including the Broadway Theater in New York, which during that period forced outside air over ice blocks placed on wooden troughs. At a Scranton, Pennsylvania, high school auditorium in 1901, comfort cooling for 1400 students was provided by 3,000,000 cubic feet of air per hour passed over blocks of ice. About 6.5 tons of ice was used to maintain a temperature of 15°F below the warm, outside air. Keith's theater in Philadelphia in 1903 reportedly would use a ton of ice per performance. And at an Indianapolis theater, the outside air of 85°F was taken into a fan through a large duct in which were placed a number of wire baskets filled with crushed ice. The system lowered the temperature in the auditorium to 70°F, but it kept four icemen busy hauling ice to the building as fast as they could.

There were, however, fundamental problems with ice cooling, recognized by many at the time. The air was cooler, yes, but it was also of uncomfortably high humidity. When warm air passes over a cold surface (such as ice), it becomes thoroughly saturated with moisture. When this air enters a hall of warm air already near saturation, it will make a fog or mist and deposit moisture on the walls. To counter this effect, some installations added pans of calcium chloride (salt) in the air stream to absorb moisture, but this practice was rarely effective because early cooling systems did not properly mix and distribute the air. A further complication of ice cooling was that it required both a willingness and proper facilities to store it on site, or to arrange for instant delivery of ice when the weather demanded it. Another handicap was the sheer cost of large quantities of ice, and at this time, many sources of natural ice were becoming polluted, resulting in foul smells.[5] All these problems could, and would be, resolved when effective mechanical refrigeration systems were developed and implemented.

Already, specialized engineers were working furiously on these very problems. In 1894, the American Society of Heating and Ventilating Engineers (ASHVE) was formed.[6] Invitations were sent to 157 individuals, and 75 agreed to join. Members were dedicated to "supplying occupants of buildings with heat to keep their bodies warm during cold weather, and with fresh air for the preservation of their life and health."[7] They obviously also regarded cooling as part of their duty to keep occupants comfortable during the hot summer months, and would soon be discussing the relationship of humidity to temperature, the means of measuring it, and the control of it.[8] This why today, we still refer to our domestic, year-round climate control systems as "HVAC" systems: heating, ventilating, and air conditioning.

By 1894 mechanical air conditioning was in the process of being transformed from trial and error methods by inventors to a genuine science. Hermann Reitschel, a professor at the Berlin Royal Institute of Technology, published the first edition of his *Guide to Calculating and Design of Ventilating and Heating Installations.* He wanted to present his specialty science in a simple and clear manner so it would be useful to designers and contractors for practical applications. His handbook marked the transfer of the science of heating and ventilation from tinkerers and scientists to pragmatic engineers, architects, and contractors. Reitschel's book included a chapter called "Kuhlung Geschlossener Raume" (Cooling of Rooms), which was possibly the first comprehensive scientific text on the subject. Before the turn of the century, Reitschel would publish a step-by-step method for calculation of cooling plants, making it possible for an engineer to design a system to condition a room for a pre-determined temperature and humidity level.[9] Reitschel's handbook would be translated and presented to the American ASHVE organization at its conference in 1896. Soon, the control of humidity became a primary objective of heating and ventilating engineers.[10]

In America, there were a number of different cooling systems being explored during this period. The first air-conditioned house in the United States was in San Lorenzo, California, by M. Dillenberg of San Francisco, in 1892.[11] In 1895, the library of a home in St. Louis was cooled by the installation of a two-ton[12] ammonia system. A ton is the industry's measure of cooling or heating capacity; the amount of energy required to melt or freeze one ton of water in 24 hours. The standard was established by the American Society of Mechanical Engineers in 1893. In 1896 an ammonia brine chiller was proposed to cool the U.S. Senate chamber. Ice-type cooling systems, that is, with air blowing over natural ice, were often used in movie theaters during the 1890s in New York, as mentioned earlier, and in 1900, mechanical cooling plants were installed on five Mississippi River steam tenders and on one steam dredge.[13]

Some heating and ventilating engineers were also exploring systems that did not mechanically cool the air but simply filtered and cleaned it. The first manufacturer and installation of a modern spray-type air washer apparently was a Thomas Acme Air Purifying and Cooling System air washer at the Chicago Public Library in 1900. An air washer cleaned and purified the air to remove particles of dirt and dust. Thomas also made a similar installation at the Chicago Telephone Company, and the result was reported in *Domestic Engineering* magazine:

> A representative of the Chronicle visited this new basement laundry the other day and was shown to the engineer. Kneeling down and opening the door where the air is drenched with water, the engineer reached a hand in toward the lower tier of pipes into which the damp air passes. The hand disappeared in the spray, and when it came out a moment later it was full of soft, watery mud of a bluish gray color. "There! That is what we get out of Chicago air," said the engineer. "How would you like to have your lungs lined on the inside with that? We get so much of it out of the air that we carry it away in buckets. On an average about one bucket of dirt is washed out of the air every day.... An attempt has never before been made in Chicago, it is said, to wash air or purify it, so as yet the atmosphere-washing at Franklin and Washington Streets is as great a curiosity as a museum freak."[14]

In 1902 the first office building to install a similar system was the Armour Building in Kansas City, Missouri. The system, installed by the B.F. Sturtevant Company, featured an air washer that went one step further and additionally cooled the air in summer by using chilled water. Each room was individually controlled with a thermostat that operated dampers in the ductwork, making it also the first office building to incorporate individual "zone" control of separate rooms.[15]

The first genuine mechanical air-conditioning projects were those designed by U.S. heating and ventilating engineer Alfred Wolff of Hoboken, New Jersey, for the board-room of the New York Stock Exchange in 1902. Wolff was probably well aware of Reitschel's groundbreaking work of 1894 in Germany. Now, a "board-room" doesn't sound large, but it actually was the special room where all the stock transactions occurred; a huge, five-story high space with dozens of circular stock-exchange bartering counters. Wolff had consulted with Henry Torrance, Jr., of the Carbondale Machine Company on the refrigeration system. One of the key objectives was to control the humidity as well as the temperature. The original system was intended to have three 150-ton ammonia absorption chillers to provide cooling for the cogeneration-type system, which was designed to control both humidity and temperature, but just before installation the system was reduced to 300 tons. "Cogeneration" means that the exhaust from the steam engines that operated the electrical generators also powered the refrigeration machines, which meant that the cooling system was powered at essentially no additional cost. The wastewater from the refrigeration condensers was stored in roof cisterns and was then used to flush the toilets in the building. Forty-two distribution boxes provided air through numerous and inconspicuous small openings in the ornate ceiling of the board-room, resulting in a downward movement of conditioned air at low velocity during the summer months. The system lowered the outside temperature from 85°F to 75°F inside, and lowered the humidity from 85 percent to 55 percent for 3,570,000 cubic feet of air per hour. It remained in operation for 20 years and was significant because it was perhaps the earliest to recognize humidity control as a primary component of air conditioning. Reduced humidity is very important to human comfort.[16] Wolff told the Stock Exchange building committee: "I would like to say that the importance of this plan ... is in the abstraction [removal] of the moisture and the reduction of humidity, I attach less importance to reduction of temperature than to the abstraction of the moisture."[17] Wolff also designed similar systems at the Hanover National Bank and the New York Metropolitan Museum of Art.

At the 1904 Louisiana Purchase Exposition in St. Louis, the public became aware of air conditioning, where they enjoyed the air-conditioned Missouri State Building along with other technological wonders.[18] The installation was for a rotunda and a 1,000-seat auditorium, cooled during the summer with about 35,000 cfm (cubic feet per minute) of partially recirculated air, cooled by direct expansion and delivered through mid-height wall registers. Refrigeration engineer Gardner T. Voorhees, who had cooled his Boston office for many years, was initially in charge, but the fair administration broke their contract and replaced him.

In 1905 Voorhees patented the multiple-effect compressor he designed for the building (U.S. Patent 793,864). The Missouri State Building undoubtedly impressed thousands of fair visitors, many of who had never experienced comfort cooling before. The building was the largest of the state buildings; however, it burned the night of November 18–19, just eleven days before the fair was to end. Most of the interior was destroyed, and it was not rebuilt. But the Missouri Building also caused a big stir among refrigeration engineers, causing *Ice and Refrigeration* magazine, which began publication in 1891, to accurately envision the future:

> The practical application of mechanical refrigeration to air cooling for the purposes of personal comfort, no doubt has a field, ... and the day is at hand, or soon will be, when the modern office building, factory, church theater and even residences will be incomplete without a mechanical air cooling plant.

Their prediction proved to be correct, of course.

Also at the fair, a self-contained mechanical refrigerator was on display by the Brunswick Refrigerating Company, which specialized in designing small refrigerators for residences and butcher shops. The ammonia system was mounted on the side of a wooden, icebox-type refrigerator. The science of air conditioning had begun to evolve, comfort cooling had been publicly displayed, and the heating, ventilating, and refrigeration industries were well developed commercially by this time. But it would take an individual who had the brains and the vision to pull all these things together into one cohesive concept and enthusiastically sell it to the public.[19]

This individual was engineer Willis Haviland Carrier (1876–1950), who is the one often credited with the invention of modern air conditioning, although, like all modern technologies, it had evolved over time with the dedicated efforts of many contributors. In 1901, Carrier, a recent graduate of Cornell University's School of Engineering, was working as research director for the Buffalo Forge Company. Carrier had responded to a quality problem experienced by the Sackett-Wilhelms Lithographing and Publishing Company in Brooklyn, New York. The hot temperature and humidity during summer days caused paper to curl and printing to become misaligned. After scientific investigation in his research laboratory, Carrier's solution was a refrigeration machine that blew air over cold coils to control room temperature and humidity. The basic function of an air conditioner, of course, is to control temperature, control humidity, control air circulation, and

Willis Carrier (1876–1950).

cleanse the air, just as Gorrie had envisioned for hospitals in 1844. It did not go unnoticed that workers in the air-conditioned printing company environment were more productive, with significantly lower absentee rates. "Comfort cooling," as it became known, appeared to be a profitable commodity in itself.

After several more years of refinement and field-testing, in 1906 Carrier was granted U.S. Patent 808,897, which he called *Apparatus for Treating Air*, and is also called his "Dew Point Control" system, the world's first refrigerated, spray-type air conditioning equipment to control humidity, as covered in his 1907 U.S. Patent 854,270, *Method of Heating and Humidifying Air*.[20] The Buffalo Forge Company began manufacture of these machines in 1905 as "Buffalo Air Washers."[21] The machines were Carrier's perfection of the primitive evaporative coolers developed in France in 1840 by Eugène Péclet. Carrier later established the Carrier Air Conditioning Company of America, a subsidiary of the Buffalo Forge Company. Carrier had seized upon the words "air conditioner" for his company name after it had be used by another inventor. This was Stuart Cramer, a textile mill engineer in North Carolina that first called this process "air conditioning" in his patent filed in April 1906 (U.S. Patent 852,823). Cramer later told how he arrived at the name: "When entering this field, several years ago, I was puzzled to find a word that would embrace the whole subject; in casting about, I finally hit upon the compound word, 'Air Conditioning,' which seems to have been a happy enough choice to have been generally adopted."[22] Cramer's patent created a ventilating device that added water vapor to the air of textile plants. The humidity made yarn easier to spin and less likely to break.[23]

The first office building specifically designed for air conditioning was Frank Lloyd Wright's (1867–1959) Larkin Administration Building in Buffalo, New York. Completed in 1906, it was described in detail by British architectural critic and writer Reyner Banham (1922–1998):

> The internal atmosphere was serviced as follows: air from well above the external pollutants was draw down capacious ducts in the blank walls of the corner towers, at the sides of the staircases, as Wright indicates. In the basement it was cleaned and heated, or after the installation of the Kroeschell refrigerating plant in 1909, cooled—but never humidity controlled, and hence, Wright's judicious quotation-marks around the words "air-conditioned" (in the town where Carrier was perfecting humidity control, he had better be careful!). [Kroeschell used safe, non-flammable carbon dioxide as a refrigerant.] The tempered air was then blown up riser ducts in the massive blank brick panels on the exterior walls immediately adjacent to the stair towers, and distributed floor by floor through input

> registers on the backs of the downstand beams under the balustrades of the balconies. The same blank brick panels also contain the exhaust ducts through which vitiated air is extracted, and a third duct-space in the panel houses pipes and wiring and other ancillaries. Throughout the interior, extracts are marked by a characteristic and much-used Wrightian device—a pattern of hollow bricks forming a coarse grille, around which even the earliest photographs show the typical staining that commonly marks an exhaust.[24]

The first air-conditioned hospital, the Boston Floating Hospital, was also completed in 1906, and was redesigned in 1907 by Edward Williams of the consulting engineering firm Westerberg and Williams to maintain hospital wards at 70°F and 50 percent humidity. Air was cooled and dehumidified using ammonia-refrigerated brine coils. The air supply was then reheated with a steam coil. Thermostats individually controlled the five wards. The system featured "reheat" in which cooled air was heated slightly to lower its humidity.[25]

Early applications of true mechanical air conditioning were mostly for large public buildings or factories. As described earlier, many of these, especially movie theaters, currently were still using crude ice-cooling systems, which were cumbersome and costly to operate and were plagued with high humidity. The new mechanical systems would replace these and other public buildings in a relatively short period of time. Within 15 years, air conditioning would become a familiar luxury enjoyed by millions of moviegoers.

Some of this transformation was the result of the innovations of Frederick Wittenmeier and Leo Logan of the Carrier organization. In 1915, Carrier and several partners had formed the Carrier Engineering Corporation, dedicated to improving the technology of air conditioning.[26] Frederick Wittenmeier was a German immigrant who had joined the Kroeschell Brothers Ice Machine Company in 1896 and had urged the company to enter the expanding ice machine business. In 1907 he designed and installed mechanical air-conditioning equipment at the Congress Hotel in Chicago. The Folies-Bergère Theater in New York installed a mechanical system in 1911 designed by Walter Fleisher but used a Thomas air washer with no mechanical refrigeration. It was not very good, as it cooled the air only seven degrees below the outside air temperature. That same year, the Orpheum Theater in Los Angeles installed a refrigerating system using a carbon dioxide direct-expansion refrigerating system made by Kroeschell Brothers Ice Machine Company, probably designed by Frederick Wittenmeier.[27]

By 1917, dozens of movie theaters were comfort cooled, with marquees proudly proclaiming, "It's 20 degrees cooler inside." And indeed, it was. Among them was the New Empire Theater in Montgomery, Alabama. Wittenmeier in 1919 and 1920 designed refrigeration systems for several Balaban and Katz theaters—the Central Park and the Riviera, in Chicago. Both cooling systems used direct-expansion carbon dioxide systems, but both proved to be less than satisfactory because cool air was supplied at the floor level. Patrons complained of drafts, often resorting to wrapping newspapers around their feet.[28] Most early theater systems were originally designed in order to optimize heating, and to do so, used upward air distribution. When they were retrofitted with cooling systems, cold feet resulted.

Europeans already knew of this problem. As early as 1864–1867, French physicist Arthur Jules Morin (1795–1880) had mentioned comfort cooling at the Conservatoire Impérial des Arte et Métiers (Imperial Conservatory of Art and Craft) in Paris, saying that upward distribution of air resulted in "the presence of very disagreeable draughts about the legs of the people, showing the mistake of letting the fresh air enter at the floor."[29] It soon became clear that downward distribution would avoid the discomforts caused by upward distribution. In 1914, a translated article in *The Heating and Ventilating Magazine* discussed the advantages of downward distribution as applied in Europe.[30] In addition, it was well known by many American ventilating engineers. All of Alfred Wolff's comfort cooling systems incorporated downward ventilation.[31]

Improvements would soon solve the "cold feet" problem. In 1922, Leo Logan Lewis, of Carrier Engineering Corporation designed the air conditioning system for Grauman's Metropolitan Theater in Los Angeles, the first to combine downward distribution of supply air, recirculation of some of the return air, and a bypass of some supply air around the cooling coil, later remixing it to provide better humidity control. All were controlled automatically to set a new comfort standard for movie theaters. Of course, the refrigeration equipment was immense, was located in the ceiling, and weighed tons, and the cost of installation was incredibly expensive.[32]

During the 1920s, the increasing popularity of moving pictures would cause an explosion of theater construction and renovations, and many demanded the installation of effective air-conditioning systems. While four theaters were cooled in Chicago in 1922, fourteen were in operation in 1925, and it was projected that fifty theaters in New York would be cooled by 1927, the year when "talkies" were introduced.[33]

Meanwhile, the use of ice in domestic iceboxes continued to grow.

With both mechanical and natural ice sources now at maximum production levels, and with constantly increasing public demand, the sale of home iceboxes doubled from 1909 to 1919. In Philadelphia, Baltimore, and Chicago, over five times as much natural ice was used in 1914 as in 1880. In New Orleans, the increase was thirteen-fold.[34] At its peak in 1900, an estimated 90,000 people and 25,000 horses were involved in the ice trade capitalized at $28 million ($801 million in today's dollars) and using ice houses capable of storing 250,000 tons each.

Plant ice production in New York also doubled from 1900 to 1910. By 1914, more plant ice was being produced in the United States each year (26 million tons) than naturally harvested ice (24 million tons). There was a similar trend around the world. Britain, for example, had 103 ice plants by 1900, making it increasingly unprofitable to export ice from the United States; annual ice exports fell to 15,000 tons by 1910.[35] The future of the manufactured ice business seemed secure and sustainable. A day would come when the natural ice business would be essentially replaced by the manufactured ice business.

At the turn of the century, millions of households were still using ice in their trusty iceboxes. But some could see the day when no ice would be needed at all for domestic refrigeration. Engineers began thinking for the first time about small mechanical refrigerating machines that could be located in shops, restaurants, and private homes. They considered alternatives of pipeline plants and individual installations, the latter being favored. This development would result in the domestic electric refrigerator, refrigeration science's greatest contribution of the 20th century. People would make their own ice and refrigerate food in mechanical refrigerators without needing to buy any ice at all.

However, small refrigerating machines could not be simply a downsizing of the huge and heavy commercial ice machines. Such machines required constant attention and adjustments by technical operators, required considerable space, and were far beyond the financial reach of most homeowners. A new system, a new approach, with automatic controls, had to be developed, primarily by inventors rather than by the existing commercial refrigerating machine companies, because these companies had more commercial business than they could handle. Only one of these commercial companies, the De La Vergne Refrigerating Machine Company, had tried to explore the house market in 1893. It "was one of the first companies to experiment with such machines, and sold a number of them, but the business did not prove to be profitable, too much personal attention being demanded."[36]

An early pioneer in this new domestic field of refrigeration was Fred Kimball, manager of the fractional horsepower electric motor division of General Electric (GE), who saw enormous sales potential in small household systems. In 1902, he convinced a group of investors to acquire several companies that were experimenting with such systems, and he formed the Federal Automatic Refrigeration Company of Cleveland. A consultant was hired to study current patents and to design a system combining the best of various options. The result was a refrigerator that "controls the operation of the [electric] motor (or gasoline engine)-driven compressor" that regulated other automatic devices. Sizes from one to ten horsepower were manufactured that produced the equivalent of from 800 to 1,200 pounds of ice per day.[37]

Other similar companies made light commercial machines of similar size, which were still relatively large, but very few were used in private homes. In 1909, there were 1,400 refrigerating plants installed, but only 5 were listed for residences, presumably among the well-to-do.[38] Many of these were "separated" machines; that is, the refrigerating machinery was sold separately from the refrigerating compartment, which may have well been the existing icebox. The machinery was set up next to the icebox, or in the basement, while the icebox remained in the kitchen. It was a relief to the consumer to have the noise, dirt, and oil of the machine in a remote area, but many of the machines were water-cooled, not so good when water pipes froze in winter or when water leaked into parts of the machinery.[39]

Many of these early household machines were unreliable and needed considerable service attention. The functional problems were many: leaky seals, refrigerants that were toxic, impure refrigerants, unreliable electric motors, lack of adequate service, ineffective automatic controls, inadequate insulation, rot developing in wooden cabinets, the inability to accurately control the small amounts of liquid refrigerants, and of course the relatively high cost.[40] One of the first companies to produce a completely self-contained refrigerator was the Brunswick Refrigerating Company, one of the few that were successful in this early household market. It housed the condensing unit in a compartment on the side of the refrigerator and was based on patents by Richard Whitaker issued between 1905 and 1912. As mentioned earlier, he had displayed his machine at the 1904 World's Fair in St. Louis.[41]

Then there was the problem that few private homes had electricity. Electrification of private homes was a slow process, due to the cost and difficulty of building the enormous infrastructure of substations and delivering electricity to private homes with power lines, often done through local cooperatives with government loans. In 1907, only 8 percent of homes were wired for electricity, mostly urban. Further, some cooperatives produced direct current (DC) electricity, while others used alternating current (AC), so manufacturers of any electrical equipment had to produce two different models, to match the local standard current. It was not until 1910 that electric current was standardized nationally at 110–115 volts AC. This simplification would enable the use of electricity to double to 16 percent by 1912, and to double again (to 35 percent) by 1920. It was not until the mid–1930s that 90 percent of urban areas had electricity, and even then, only 10 percent of rural areas had it.

Probably the earliest manufacturer of consumer electrical appliances was the General Electric Company (GE), formed in 1892 by the merger of the Edison General Electric Company and the Thompson-Houston Company. At that time, GE invested heavily in AC power. One of the predecessor companies, the Edison Electric Light Company, had begun the electrical industry in the United States in 1879. The first household consumer electric appliance by GE was the ubiquitous electric fan, made and marketed in the 1890s, followed by the electric iron in 1905. By 1907, GE's appliance line included hair dryers, electric home heaters and cooking devices such as teakettles, cookers, toasters, and coffeemakers. In 1908, an electric vacuum cleaner, known as the Invincible Electric Renovator, appeared on the market by the Electric Renovator Manufacturing Company of Pittsburgh, using GE's motor and electrical equipment. In 1911, the first electric range was manufactured by the Hughes Electric Heating Company of Chicago, which would in 1918 merge with Hotpoint and the heating device section of GE to form the Edison Electric Appliance Company to produce Hotpoint brand products. By 1934, Hotpoint and GE brands would be integrated within the GE organization.

In 1911, GE agreed to manufacture a commercial refrigerator for the Audiffren Company, which held the rights to an 1895 U.S. patent, number 551,107, granted to a French Cistercian Monk and physics teacher, the Abbé Marcel Antoine Audiffren, who was a teacher at the Petite Séminaire in Grasse, France, in 1904. He had developed a hermetically sealed refrigerating system to cool the wine made by the monks. Like the Romans and Greeks eight centuries before, modern man still liked his wine cooled. The system incorporated a stationary compressor within a rotating bronze shell, which was operated by a hand crank. Compressed sulfur dioxide refrigerant condensed on the outside of the shell, which rotated within a water bath. The condensed liquid, held against the shell by centrifugal force, was skimmed off, then passed through a wall to another rotating shell, also rotating in a water bath. The liquid refrigerant evaporated as heat transferred from the water being cooled through the bronze shell, and the vapor passed back through a shaft to the first shell, where it was again compressed.[42]

The Audiffren machine had been commercially produced in 1904 through the efforts of French industrialist Henri Albert Singrun in his Singrun Company at Epinal, France, and later in other countries. It was patented in 1908 (U.S. Patent 898,400, *Rotary Refrigerating Machine*). Some years later, an enterprising American international trader, Griscom, was touring France and while in Epinal, became convinced that the Audiffren-Singrun (A-S) machine had commercial possibilities and obtained rights to build it in the United States. Initially, sales were though the H.W. Johns-Manville Company, which had an extensive marketing organization, but it soon was contracted out to the Fort Wayne Electric Works division of General Electric Company, which started manufacture of it in 1912. James T. Wood, manager of the Fort Wayne facility, saw the A-S as a boon to the sale of electricity. The A-S machine was made of wood, similar to current iceboxes, but larger, about the size of a bedroom wardrobe with three

The Abbé Marcel Antoine Audiffren ("Audiffren Drinking Water System," Bulletin D-201, 1921, H.W. Johns-Manville Company).

Opposite: ***Audiffren's 1895 U.S. Patent No. 551,107. A stationary compressor was mounted in a bronze shell at right. Compressed sulfur dioxide refrigerant condensed on the outside of the shell, which was rotated in a water bath. The condensed liquid held against the shell by a centrifugal force was skimmed off, then passed through the shell at left, rotating in the water to be cooled. The liquid refrigerant evaporated as heat transferred from the water being cooled through the bronze shell. The refrigerant vapor thus formed then passed through the shaft, to the right shell, and into the compressor.***

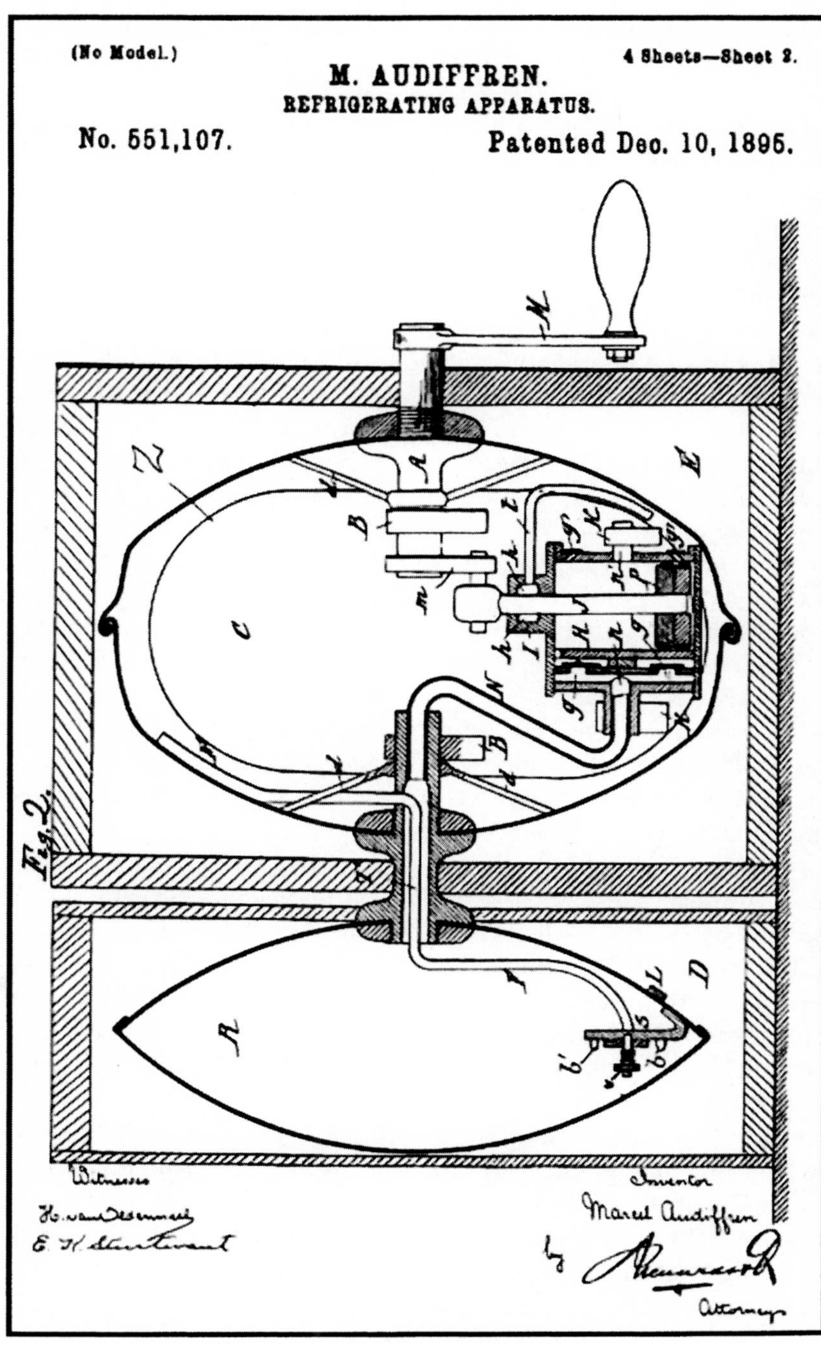
(No Model.)
4 Sheets—Sheet 2.
M. AUDIFFREN.
REFRIGERATING APPARATUS.
No. 551,107.
Patented Dec. 10, 1895.
Fig. 2.
Witnesses
Inventor
Marcel Audiffren
Attorneys

doors, and had the refrigeration equipment mounted on top, concealed by a drop-down door for service access. GE would produce the A-S until 1928 in small quantities, but it was bulky and expensive, was not really suitable for household use, and cost $1,000, twice as much as a new car.[43] However, in improved versions, it was durable, and it would inspire GE's intense interest and research in refrigeration and lead to GE's later successful commercial developments. The A-S machine would be manufactured into the 1940s by license from Singrun, and some say that there is a possibility that A-S machines are still in operation in underdeveloped countries, where many were sold.[44]

Possibly the first lightweight, compact, and inexpensive household electric refrigerating machine, was the "DOMELRE" household refrigerator introduced in Chicago in 1913 by Fred W. Wolf, Jr., son of a brewery architect and ammonia refrigeration engineer. Patents were filed in 1913 and granted in 1920 as a *Process of and Apparatus for Refrigerating*. The name "DOMELRE" was a contraction of "DOMestic ELectric REfrigerator." The entire unit was self-contained, with the condensing unit mounted on a wooden base with the evaporator hung underneath. The entire assembly could be placed on top of an existing icebox in which a hole had been cut to admit the evaporator. An air-cooled condenser made of bare copper tubing was used, and a ¼-hp repulsion-start electric induction motor powered the machine. The refrigerator included an ice cube tray, the first household unit to feature one as part of the evaporator, and the refrigerant used in the evaporator was wet sulfur dioxide. The machine operated automatically with a thermostat.

The low current that was required permitted the unit to be connected into an ordinary Edison-base-type electric lightbulb socket with a flexible cord ending in a threaded connector, similar to a lightbulb, to screw into the socket. In 1913, appliances were a totally new technology and homes were not wired with electrical outlets, other than for lightbulbs, so early appliances had a threaded male "Edison base" as used on lightbulbs, which was simply screwed into an available empty light socket. Kitchens of the day became nightmares of such "octopus wiring." Multiple-socket adaptors were screwed into a ceiling drop-cord light socket in the middle of a kitchen, and multiple cords from appliances were screwed into the adaptors, trailing through the air to their respective appliances.[45]

The DOMELRE was made by the Mechanical Refrigerator Company in Chicago but did not have extensive sales until Henry B. Joy of the Packard Motor Car Company purchased the rights from Wolf for $225,000 in 1916, formed a company called Isko Incorporated, and moved production to Detroit. It was redesigned with a finned-tube, air-cooled condenser, probably the first in the industry, which was an idea borrowed from the finned radiator used on the contemporary Pierce-Arrow automobile.[46] The cost was also reduced from $385 in 1916 to $275 ($6,020 in today's dollars), installed in any icebox.

Perhaps Henry Joy was inspired by his automotive competitor and namesake, Henry Ford, who had become famous for his 1913 Model T Ford, which demonstrated the marvel of mass production and at the same time, provided thousands of consumers with a product that was affordable and reliable (the Model T cost $440). Nevertheless, various technical problems and low public demand for the DOMELRE caused Joy to lose interest, and he sold the company. The new owner redesigned the machine, which caused further problems, and production soon ceased. Eventually, Frigidaire would purchase the company for $17,400 in 1922 to acquire the patents.[47]

The prospect of creating the very small, reliable, refrigerators needed for home use still looked dim as late as 1916. At that time, John Starr, the first president of the American Society of Refrigerating Engineers, was not encouraging:

> That "philosopher's stone" of the refrigerating engineer—"the domestic refrigerating machine that could be run by the cook"—was just as active in the minds of the men of 1891 as it is today and the amount of money and thought spent on this subject was perhaps no less than it is now. The small machine requiring some intelligent, though not highly skilled attention, has become an ... acceptable and widely used apparatus ... [but] that supposed line of demarcation between a machine that required a little attention and no attention at all, was found not to be a line but a gulf which has perhaps not yet been successfully bridged.[48]

The machine for the home had to be idiot-proof to be successful in the market, and at this time, that was only an engineer's dream.

In about 1913, Nathaniel B. Wales, an automotive device inventor, had teamed up with Detroit engineer Edmund J. Copeland, a former general purchasing agent for GM. They wanted to develop an absorption-type household refrigerator that Wales had conceived. It used condensation equipment housed in the basement. GM president William Durant warned Copeland that he was a fool to put his money into such a hare-brained idea, but the two proceeded anyway. Copeland and Wales obtained funding from Arnold H. Goss, director of the Chevrolet Motor Car Company. A prototype was subsequently constructed, but Goss rejected it as impractical.

Copeland and Wales in September 1914[49] started working on a vapor-compression machine. Also in 1914, Goss and Copeland began an association with the Grand Rapids Refrigerator Company, founded by Charles, Frank and Fred H. Leonard, who since 1881 had been manufacturing a line of wooden iceboxes variously named "Leonard Cleanable," "Northern Light," and "Challenge." In 1901, some were made under the brand of the Northern Refrigerator Company of Grand Rapids.[50]

The Leonard "Cleanable" came about because of a mishap in the Leonard home. A pail of hot lard was left inside the icebox on top of a cake of ice, resulting in melted ice, a spilled pail, and cooled lard spilled all over. Charles created his refrigerator with removable liners and flues to avoid such a problem. In 1885, Leonard introduced metal shelves and improved the door-locking mechanism. The introduction of porcelain-lined interiors in 1909 improved ease of cleaning. The refrigerator cabinets were made of highly varnished carved oak with brass fixtures and were enhanced with mirrors.[51]

Apparently, the Leonards were interested in electrifying their product, and Goss and Copeland were interested in Leonard's construction of wooden cabinets for their refrigeration mechanisms. In May 1916, the company incorporated as Goss and Copeland Electro-Automatic Refrigerator Company. A few months later, the company name was changed to Kelvinator and was run by Copeland.[52] By 1917, after a dozen different models had been constructed, a machine that worked "reasonably well" was perfected.[53] It was made of wood and was probably the first electric refrigerator that made it to production. Copeland had developed an automatic control device and a solution to the problem of gas leakage.[54] By that time, Wales had departed and Fred Heideman, an assistant of Fred Wolf, Jr., had come over from Isko. By February 1918 Kelvinator installed a final prototype in a home as a field test, which was successful, and the system, which had some features of the DOMELRE/Isko, was placed on the market. Sixty-seven machines were built and sold that year. Some of these were under the Leonard brand name and others under Kelvinator's.[55] Customers bought an icebox for the kitchen and Kelvinator installed its condensing unit (called the "bird cage") in the basement, and connected the two parts with a series of pipes that went through a hole in the floor. Kelvinator would not combine the machinery and the refrigerator until 1925. In 1926, the companies would merge and would be known as the Kelvinator-Leonard Company.

In about 1916, William Durant of GM apparently had a change of heart about household refrigeration. He attempted to buy Kelvinator, whose founding he had earlier discouraged. When he was unable to do so, he purchased a bankrupt Detroit firm, the Guardian Refrigerator Company, in 1918. Guardian had been founded in 1915 by Alfred Mellowes and Rueben Bechtold and was one of the many small companies that had failed with household refrigeration. The company had tried to market a self-contained electrical refrigerator (U.S. Patent 1,276,612) in 1916,[56] but had sold fewer than three dozen and had lost $40,000 trying.[57]

Apparently, Durant wanted to have an alternative source of income should the automotive industry be ordered to cease automotive production during World War I, then in progress, and thought that refrigerators could tide his car dealers over during the war emergency. But his dealers were not as enthusiastic as Durant was for the new product, which had forced GM to set up separate dealerships just for refrigerators.[58] When asked why he had bought Guardian, Durant replied, "What are refrigerators but boxes with motors?" pointing out that GM was already in the business of making boxes with motors.[59] GM, during the last year of World War I, 1918, had produced 246,834 "boxes with motors," mainly cars, trucks and tractors, mostly for the military. Durant renamed Guardian as Frigidaire in 1919. By 1920, Frigidaire had produced thousands of refrigerators; however, customer complaints of malfunctions poured in by the hundreds. Discouraged, Durant ordered GM vice president John L. Pratt to investigate Frigidaire with the thought, shared by other GM executives, that it might be shut down[60] because the company was losing $2.5 million per year on it.[61]

Pratt found many engineering problems, but when he talked to users, he discovered that they were fiercely loyal to the product. One woman was willing to give up anything in her kitchen except her Frigidaire. He also found that there was inadequate service for refrigerator breakdowns. He advised GM that if Frigidaire were shut down, it would leave many customers without any service at all and convinced GM that Frigidaire was a basically sound product that could be made profitable.

So GM decided to transfer Frigidaire manufacturing from Detroit, Michigan, to Dayton, Ohio, to be operated by GM's Delco Light Company division, near the recently established GM Research Laboratory. Frigidaire was in debt to GM for $3.5 million, and there was little hope for a successful product.[62] Further, in 1921 only 5,000 mechanical refrigerators were sold in the United States.[63] But three research engineers were assigned by Frigidaire in 1921 to improve the product, which they successfully accomplished after shutting down production for several months. The result was the 1921 B-9 model, in which the refrigeration machine

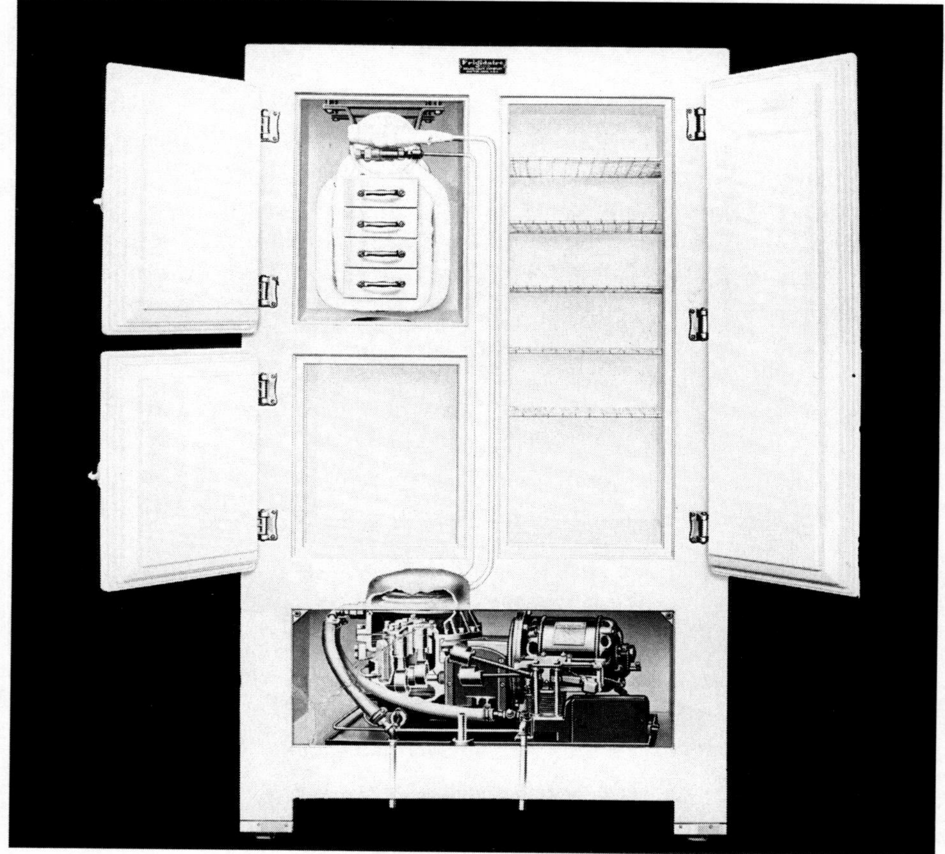

Frigidaire Model B-9, 1921.

and motor were concealed on the bottom below the food compartment. It was probably the first modern-looking refrigerator, with a plain, white-painted, all-wood enclosure. Wooden cabinets would be used until 1926. Insulation at first was seaweed and later, corkboard.

The direct expansion evaporator used a "low-side" float expansion device connected to a water-cooled sulfur dioxide condensing unit. It had four ice cube trays and adjustable shelves. The B-9 weighed 834 pounds and cost $775 ($10,100 in today's dollars). Production was resumed in limited quantities and by 1923 sales had boomed. The profits from Frigidaire were so large that they covered the entire dividend of GM preferred stock that year.[64] By 1924, Frigidaire and the Nizer Laboratories Company for the Arctic Ice Cream Company of Detroit had adapted household refrigeration to ice cream dipping cabinets, which previously had used a mixture of salt and ice. Early electric cabinets used flooded-type refrigeration systems to maintain alcohol-based antifreeze, which surrounded wells holding the ice cream cans. Small refrigerating machines were also applied to vending machines for cold water or soft drinks.[65]

The large refrigeration companies had been slow to enter the household market, and even the American Society of Refrigerating Engineers published just a few papers about such machines. In 1923, *Ice and Refrigeration* magazine decided to update its 1916 domestic refrigerating machine article to cover the various types that were then on the market:

> In the interim ... there has been considerable activity in the development and production of a successful household refrigerating machine. Some of the best engineering talent in the country has been employed to design and perfect these small mechanical refrigerating machines; others are in the experimental stages; and new devices, processes and machines are being brought out from time to time.[66]

This "considerable activity" was pretty much restricted to only a few participants. By 1924, four manufacturers controlled 90 percent of the household electric refrigerator market: Frigidaire, Kelvinator, Copeland, and Servel.[67] Almost a third of all refrigerators were sold by electric utility companies in impromptu appliance stores, and the rest by manufacturer's door-to-door salesmen. The next chapter will describe how General Electric entered the field and developed the first commercially successful household refrigerator, and the successful development of air conditioning by Willis Carrier.

Chapter 5

Gas Refrigeration and Air Conditioners

Since 1900, the use of electricity in homes had been growing. Various home electric appliances had been introduced, including clothes irons, fans, washing machines, and ranges. These had not only been promoted enthusiastically by the "central station" type of electric utilities; many of these utilities sold electrical products under their own brand names at low cost in order to increase the use of electricity. But because of the initial financial and functional problems in home electric refrigerators, the electric utilities had hesitated to promote and sell electric refrigerators. Now, in 1925, they changed their tune:

> For the past three years (1921–1924) or so, there has been a steady increase in the sale and use of domestic electric refrigerating equipment, which has now mounted to such proportions as to make it safe to assume that domestic electric refrigeration has come to stay and is not going to suffer the setbacks and almost complete demoralization it suffered on several occasions in the past, when it seemed to have received a healthy start.
>
> From the central station standpoint, almost everyone is familiar with the attractive load building possibilities of this device, resulting as it does, in a revenue representing possibly the highest rate per kilowatt of demand of anything connected to the central station system.[1]

In 1925, the managing director of the electric power trade organization in the United States, the National Electric Light Association, admitted that it was going to organize a national promotion of electric household refrigerators. And why not? Unlike most electrical appliances, refrigerators operated 24 hours per day, generating a constant stream of power income for electric companies.[2] A 1926 study revealed that simply installing an electrical refrigerator doubled the revenue received by an electric company from a home.[3] A major manufacturer of consumer electrical products was the General Electric Company (GE), which under Edison had essentially established the electric industry.

Accordingly, as a result of GE's involvement with the Audiffren refrigerator in 1911, GE's engineers at the Fort Wayne facility began experimentation with domestic refrigerating machines in 1917, hoping that a replacement for the Audiffren-Singrum (A-S) machine could be sold in greater quantities. Clark Orr, who had supervised the manufacture of the A-S machines, tested some open-drive machines but settled on a hermetically sealed compressor in 1920. Sealed within the compressor was a water-cooled coil to condense the refrigerant, making a much quieter and compact assembly. After testing and modifications, 50 samples were built, which included the oscillating cylinder design of the A-S machine, and a water-cooled condenser within the compressor shell. The condenser unit, a sort of domed cylinder about 18 inches high, was mounted on the top, and the sulfur dioxide refrigerant was housed in the bottom casing. A brine-type evaporator was used. It looked like a standard icebox with a bronze, mechanical, motor/compressor cylinder on top.

These machines worked reasonably well, but no further development action resulted until Francis Pratt, GE's vice president of engineering, asked Alexander Stevenson, Jr., of the engineering department in Schenectady, New York, to survey the entire field of domestic refrigeration. It was probably the most extensive study of engineering and economic matters of the domestic refrigerator up to that time. Stevenson identified fifty-six companies that were already in the business by 1923. Kelvinator and Frigidaire had already made several thousand refrigerators. Kelvinator had sold 75,000 in 1925, the year it combined its machinery with the refrigerator itself. Of the fifty-six companies only eight were well financed or on the way to large-scale production. Most made compression machines. The lowest price machine was $450 ($6,140 in today's dollars), a quarter of the average annual salary and about the same price as a new Model T Ford. The highest price was a 9-cubic-foot wooden model with ice cube trays, probably the Frigidaire, at $714 ($9,740 in today's dollars).

Not that many people were able to afford such prohibitive costs. In 1923 only 20,000 households in the entire United States had mechanical refrigerators (one household in 1,000).[4] Most sales were to relatively wealthy households. In 1926, 80 percent of affluent families studied in 36 American cities had vacuum cleaners and washing machines. Although people in the 1920s enjoyed reasonable economic prosperity, most middle-class people were still economically challenged. Working class families of five had to get by on $1,920 per year, as determined by a 1924 intensive study by Helen and Robert Lynd entitled *Middletown: A Study in American Culture*. Middletown was actually Muncie, Indiana. The Lynds found that half the households in town did not meet minimum standards of health and decency. Among working class wives, 55 of 124 worked for wages.

This was a new phenomenon. World War I had forced many women into the work force for the first time, while their husbands were in the service, and the practice continued after the war. This provided additional family income and the ability to purchase nonessentials such as the new electrical appliances that were being produced and promoted. Many of these were of interest to women, such as vacuum cleaners, wash machines, ranges, and toasters, which promised reduction of household drudgery for women, although a refrigerator did not mean much less manual labor than an icebox. But as a result of the 19th Amendment to the U.S. Constitution in August 1920, women were granted the right to vote. Many women interpreted this not just as a right to vote in elections, but also to vote in their own household, encouraging them to participate more forcefully in decisions about household expenditures and purchases. As a result, the mass market sought by manufacturers was becoming not only larger but more dependent on women's choices, not just men's. This trend would soon dramatically change consumer product design.

In 1924 a typical upper-middle-class urban home would have a telephone, hot and cold running water, indoor plumbing, gas and electricity, and perhaps a radio, which was another recent technological invention. But the most compelling consumer product of the time to possess was the automobile, which was the equivalent of what the computer would be in the 1990s. Even in Muncie, of 123 working families, 60 had cars. In many households, it was the car that was competing for any extra cash, because it offered freedom of travel and public exhibition of economic status. In 1924, one could buy a new Studebaker touring car for $975, just twice the cost of the cheapest electric refrigerator. And you could buy the car on time payments for just $35 per month, a practice that would soon be adopted by many appliance manufacturers, including refrigerators.[5]

But refrigerators were not yet in mass production, and because of that were still far too expensive for most middle class people. Mass production enabled manufacturers to spread the cost of manufacture over huge quantities of products, thus enabling them to significantly reduce the cost of each individual product. Stevenson's report on refrigerators for GE, begun in 1920 and completed in 1923, included the following recommendation:

> It is recommended that The General Electric Company should undertake the further development of an electric household refrigerator as an addition to their string of appliances, and because widespread adoption will

> increase the revenue of the central stations, thus indirectly benefiting the ... Company. But [the Company] should not enter this field in the hope of immediate profits from the sale.... For some years to come, the developmental and complaint expenses will probably eat up all the profits.[6]

He also wrote: "No thoroughly reliable and reasonably inexpensive domestic refrigerating has yet been developed by any of the major companies engaged in this enterprise.... During the developmental stage, it is easier to lose than make money."[7] When vice president Pratt forwarded the report to GE president Gerard Swope (1872–1957), Pratt further stated that the art of household refrigeration had reached the point of stabilization and that it was an opportune time to enter commercial competition. But he also cautioned:

GE's Gerard Swope (1872–1957) (courtesy the Museum of Innovation and Science, formerly the Schenectady Museum, Schenectady, New York).

> There reads through Mr. Stevenson's report the important fact that all existing practice carries a more than normal hazard of being revolutionized by inventions of a fundamental character, So many active minds throughout the country are being directed to the solution of these problems that it would perhaps be surprising if some such inventions did not materialize. The business is a rapidly evolving one, making real strides from the developmental to the commercial stage.[8]

GE accepted the Stevenson recommendations, but Pratt was right. Between 1923 and 1933, inventions that would profoundly alter the design of domestic refrigerators did, in fact, materialize. Up to this time, practically all of ice-making and refrigerator development had been with vapor-compression machines, all of which required steam or electric motors to power them. The major electric companies encouraged this because of the enormous profits refrigerators could bring with their continuous use of electricity. Electric companies had no interest whatsoever in any alternative technology for refrigeration. But at this time, less than 50 percent of U.S. residential homes were electrified. Therefore a huge potential market remained for refrigeration that did not require electricity. What kind of technology could provide such an alternative?

To answer this question, independent inventors looked back to the work of Ferdinand Carré in France in 1860. Carré had invented the first absorption process refrigerator. In the absorption process, as you may recall, the refrigerant is heated by a gas flame, instead of a compression machine, to cause it to vaporize. Carré's machine was of a relatively small size, suitable for households, and was exhibited at a London exhibition in 1862. Its major advantage was that it was perfectly quiet, with no moving parts and no noisy compressor with its incessant hum. But as electricity became popular near the end of the century, this technology became essentially forgotten—until 1922, that is, when in Sweden, two young engineering students, Carl G. Munters and Balzar von Platen, designed an absorption machine that would run quietly and continuously and had no moving parts. Thus it would not require expensive automatic controls or electricity. It was patented in Sweden in 1924. This machine, using the same scientific principles as the Carré machines of the 1860s, was purchased by AB Arctic, a company that would soon be bought by Electrolux-Servel. Ironically, Servel's name stood for its slogan, "SERVing ELectric," a common advertising phrase used for decades by electric companies to promote the idea that electricity was like having a servant in the home in an era when the wages of servants, mostly immigrants, were soaring out of the reach of many households. Servel had in fact been funded by a group of electric-utility holding companies to manufacture and market compressor refrigerators. But in 1925 Servel had purchased the American rights to the 1922 Swedish patents (1927 U.S. Patent 1,613,628) on the continuous absorption refrigerator and had reorganized, with five million dollars from the financial interests that controlled the Consolidated Gas Company of New York, in order to devote itself solely to gas refrigeration. Since it had a manufacturing plant already, it was able to quickly produce machines. Servel introduced a refrigerator in 1926 that was powered solely by gas heat, based on the Platen-Munters patents.[9] It was successfully marketed to rural areas that did not have electrical service.

In 1926, when the American Gas Association met in Atlantic City for its convention, there were only three manufacturers of gas refrigerators, and only one, Servel, would succeed in reaching mass production. The other two would fail within a few years. One of these was the Sorco refrigerator, the creation of Stuart Otto, an engineer who had patented an absorption refrigerator in 1923. He owned a factory in Scranton, Pennsylvania, that produced dress forms for seamstresses, and he persuaded twenty businessmen to put up $5,000 to develop his machine and adapt his plant to produce it. His units were sold to utility companies and seemed workable, so he decided in the fall of 1926 to attempt large-scale production.

Servel gas refrigerator, 1926.

Otto tried to interest General Electric and General Motors in his refrigerator, but GE was about to bring out its own refrigerator, and General Motors had just purchased patent rights for an English machine (called the Faraday) that used a solid rather than a liquid solvent and that would be marketed on a limited basis in the 1930s, although GM soon dropped it. Attempts made by Otto to raise the $1 million needed independently also failed, so Otto made an option agreement with his stockholders to buy their stock and pay them within a year. He then offered manufacturing companies non-exclusive licenses for the manufacture of his machines under the Swedish patents. He described the results: "Each of three companies paid me a cash down payment on signing of $25,000 and agreed to a guaranteed minimum of $35,000 per year royalty on 5 percent of net sales, for 17 years."[10] But within a few years, Otto had to admit defeat. He was unable to carry the financial load, and after two years he managed to collect only a small portion of the accrued royalties. His small company was unable to compete against the giants of the industry, GE and GM. Thus, Servel was essentially the only major American gas refrigerator company from 1926 to 1956, when it ceased production of refrigerators.[11]

Other independent inventors saw the introduction of the 1926 Electrolux-Servel as opening a new line of research into household refrigerators that used less electricity, were quieter, and were smaller in size. The most famous of these were Albert Einstein (1879–1955) and his former student, Hungarian scientist Leó Szilárd (1898–1964), who collaborated on ways to improve home refrigeration technology starting in 1926 in Germany.

They were motivated by contemporary newspaper reports of a Berlin family who had been killed when a seal in their refrigerator broke and leaked toxic fumes into their home. Einstein and Szilárd proposed that a device with no moving parts would eliminate the potential for seal failure. Einstein used his experience gained from his years at the Swiss Patent Office to apply for patents for their inventions in several countries, including the United States (Patent No. 1,781,541, filed in 1927), for three different models under both their names. It has been suggested that Szilárd did most of the actual inventing, and that Einstein just acted as a consultant and helped with the patenting process. Einstein was totally uninterested in business, but Szilárd was poor and the idea seemed interesting to him at the time. He thought that with Einstein's name on the idea, it would surely sell. Of course, by then, Einstein was already world famous for his theory of relativity, conceived in 1911 and proven in 1919. In 1921 he had received the Nobel Prize in physics for his explanation of the photoelectric effect, as the theory of relativity was still somewhat controversial.

Their joint patent described an absorption refrigerator without moving parts, operated at constant pressure, and requiring only a heat source (such as a small gas burner or even solar energy) to operate. It was an alternative design based on the original 1922 invention by Swedish inventors Baltzer von Platen and Carl Munters, the inventors of the 1926 Electrolux-Servel absorption refrigerator. All absorption machines are based on the absorption principle developed by Ferdinand Carré in France in 1860.

The Einstein-Szilárd machine used an ammonia pressure-equalizing fluid, butane refrigerant (not soluble in water), ammonia, and water. A water-flow loop served as an ammonia pump and the ammonia-flow loop served as a butane pump. The refrigerant liquid evaporated, mixed with a flow of ammonia vapor to form a gas, and the mixture flowed, not to a pump, but to an ammonia absorber. The absorber worked by removing ammonia vapor by dissolving it in water, allowing the refrigerant to condense and deliver heat to an external convector, as in a conventional refrigerator.

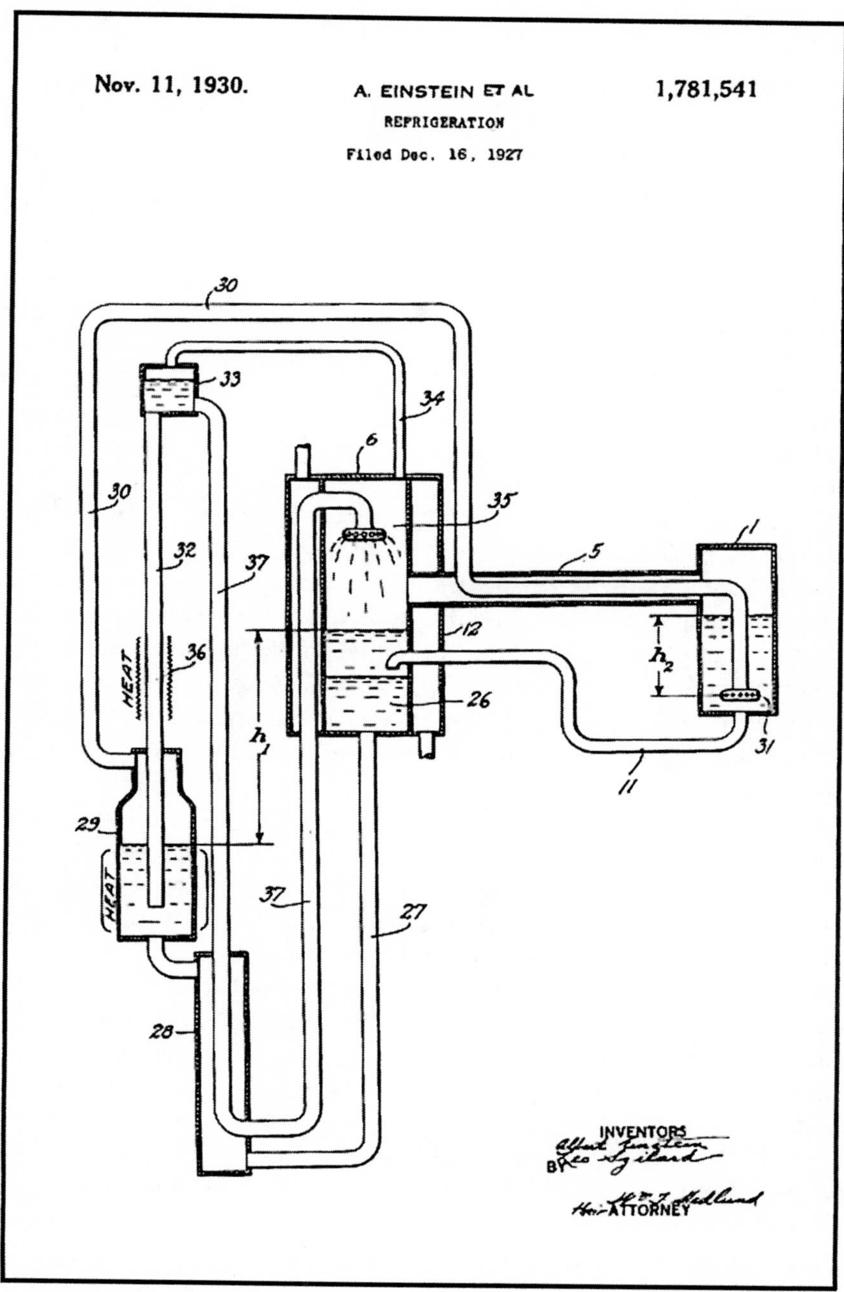

Einstein-Szilárd U.S. Patent No. 1,781,541, for refrigeration, 1927.

The refrigerator was not immediately put into production, but the Swedish company Electrolux, which had recently introduced the 1926 Electrolux-Servel absorption refrigerator, bought some of the critical patents.

Einstein was visiting the United States in February 1933, when he decided not to return to Germany due to the rise of Hitler to power. He undertook a two-month visiting professorship at the California Institute of Technology. He and his wife returned to Belgium in March, where he learned that the Nazis had raided their cottage and confiscated his personal sailboat. He immediately went to the German consulate, where he renounced his German citizenship. In April, he learned that the Nazis had passed laws barring Jews from teaching at universities. Since he was a Jew, this ended his career in Germany. After a few years in Belgium and England, in 1935 he decided to emigrate to the United States and become a citizen after accepting a position at the Institute for Advanced Study in Princeton, where he would stay until his death.

Szilárd, on the other hand, fled to London in 1933 when Hitler came to power. There, he conceived the concept of nuclear chain reaction and in 1936 patented it with the British Admiralty to insure secrecy. He was also a co-holder with Enrico Fermi (1901–1954) of a patent on the nuclear reactor (U.S. Patent 2,708,656). In 1938 he and Fermi accepted an offer to conduct research at Columbia University in Manhattan. They learned of the nuclear fission experiment by German scientists, and in 1939 Szilárd wrote a letter to President Franklin Roosevelt, explaining the possibility of nuclear weapons, warning of their development in Germany, and encouraging the development of a program that could result in the creation of an atomic bomb. He approached his old friend Albert Einstein and convinced him to sign the letter, lending his fame to the proposal. This letter initiated research into nuclear fission and ultimately led to the Manhattan Project and the atomic bomb in 1945.[12]

Meanwhile, electric refrigeration development had continued by the major companies. GE's Clark Orr at Fort Wayne had proceeded with the development of the "OC-2" (Oscillating Cylinder) vapor compression machine, which replaced the original water-cooled condenser with an air-cooled version. Twenty test units were built in late 1923 and another hundred in early 1925. The new design was announced in late 1925 as "the General Electric Refrigerator." Finned condenser tubes were wrapped around the welded-shell and a 1/6-hp electric motor/compressor was placed on top of the cabinet on the right side, allowing it to be cooled by natural ventilation. The motor/compressor assembly was a cylinder about 18 inches high and 18 inches in diameter.

The wooden cabinet was insulated with corkboard and was available with an ivory enamel or natural oak finish in 9- or 15-cubic-foot sizes, matching exactly the finishes and sizes of Frigidaire's B-9 and B-15 models, already on the market. Because the numerous iceboxes then in public use varied greatly in construction and efficiency, GE decided to sell its refrigerators as a complete unit. The cabinets were manufactured by the Seeger Refrigerator Company in St. Paul, Minnesota (GE would not fit its machines into other iceboxes), and shipped from the factory directly to the customer, as were the refrigerating units from Fort Wayne. Within a year, 2,000 had been sold by GE's Central Station Department through electric utilities.[13] Electric companies across the country were staging expositions to educate potential customers and to sell people new refrigerators.

Although a great improvement over the Audiffren-Singrun unit, GE did not feel the OC-2 was the final answer. So in late 1925, GE started a new round of development, this time in Schenectady, New York. Competing teams of engineers began the task of developing a mass-production refrigerator. The winner, a new design, the "DR-2" (DR for "domestic refrigerator") was selected after extensive testing. Christian Steenstrup (1873–1955), a gruff Danish immigrant who migrated to America in 1894, was the lead engineer of the winning team.[14] Steenstrup had started with GE in 1901 in Schenectady and at the time of the DR-2 was supervisor of mechanical research, responsible for the design and development of a variety of equipment throughout the plant.[15]

The new design was initially to be called the "Freosan," but it was feared that such an unfamiliar name would not be recognized by the public, so at the last minute before introduction in late 1927, the name was changed to simply the "Monitor Top," because the refrigeration cylinder on the top resembled the cylindrical gun turret that sat on the flat deck of the famed Civil War iron-clad battleship, the USS *Monitor*. The OC-2 had had a similar exposed refrigeration cylinder on its top, but on the right side. This time, the cylinder was centered on the top, rather than to one side. Its fundamental difference was that the cabinet was made of steel, not wood, a technology pioneered by the automotive industry that had been first used by Frigidaire in 1926. The Monitor Top's capacity was 5 to 7 cubic feet of food storage with an ice cube tray, and it was elevated above the floor about 8 inches by four furniture-type legs. It used either sulfur dioxide or methyl formate as a refrigerant. The motor was hermetically sealed to reduce noise. The DR-2 had a low consumption of electricity, only 50 kWh (kilowatts per hour), half the power used by competing refrigerators. GE spent $18 million for the manufacturing plant and $1 million for the aggressive 1927 advertising campaign. It was promoted as the "first affordable refrigerator," at $300 ($4,020 in today's dollars), and as the "Model T of refrigerators." The basic design would be continued until 1936.

In 1927 GE formed a separate electric refrigeration department, headed by Steenstrup as chief engineer and headquartered in Cleveland, Ohio. GE opted for mass distribution and sales by dealers, rather than through the electric utilities, which would benefit by the increased use of electricity. Potential revenue for utility companies was enormous, because unlike other appliances, the refrigerator ran 24 hours a day. GE was the first to sell the cabinet as an integral part of its refrigeration machinery. The Monitor Top stunned its competition with its advanced features and low price. Even more stunning was its claim in sales literature that it had created

General Electric "Monitor Top," 1927 (courtesy the Museum of Innovation and Science, formerly the Schenectady Museum, Schenectady, New York).

> an electric refrigerator so efficient, so simplified, that it could literally be installed and forgotten ... a refrigerator so simple that all you need to do is to plug it into the nearest electrical outlet and it never even needs oiling. There is not a single exposed moving part. There isn't a single belt, fan, or drain pipe—nothing below the cabinet—nothing in the basement.[16]

This contrast to previous refrigerators and iceboxes could not have been greater. It came in eight models, ranging from 2.5 cubic feet to 18 cubic feet, to fit any size kitchen.[17] In its first year it captured 7 percent of the domestic refrigeration market. By 1929, 50,000 Monitor Tops would be sold and by 1931, GE would claim a million users. GE would produce these for the next ten years, and they were so reliable that thousands are still in use today. Within a year of its introduction, most other refrigerator manufacturers were fighting back by producing steel cabinets by mass-production, instead of wooden cabinets, which had to be built by hand.[18] Frigidaire, a step ahead, was already introducing the first porcelain-on-steel refrigerator exterior, and Kelvinator was advertising: "You needn't buy a new one to enjoy Kelvinator." It offered a Kelvinator "Zone of Kelvination," a separate refrigeration unit, that could be installed into any existing icebox.

Meanwhile, Sears, Roebuck and Company had debuted its Coldspot refrigerator in 1928. It was something of a novelty, but it met with immediate public acceptance, and its 1929 model was a main point of interest at the Paris International Exposition. That year, GE introduced a combination refrigerator with a frozen food storage space that had its own evaporator and door, and a fresh food storage space with its own cooling system and separate door.[19]

Frigidaire Division of General Motors, which had dominated the household refrigerator market in the 1920s with 80 percent of the market, saw its sales steadily decline after the introduction of GE's Monitor Top. But by 1929, when the one-millionth Frigidaire was produced, even Frigidaire's president could say: "Electric refrigeration for the home has definitely passed out of the 'Maybe—some day!' stage. It is entering the 'What?—haven't you got it yet?' stage."[20]

Indeed, the sale of mechanical household refrigerators was increasing at a phenomenal pace. Total annual household refrigerator sales nationwide in 20 years were:

Year	Sales
1910	100
1912	175
1914	600
1916	1,000
1918	1,400
1920	4,000
1922	10,000
1924	17,000

Year	Sales
1925	75,000
1926	248,000
1928	468,000[21]

Electric refrigerators were promoted enthusiastically by the major manufacturers in all major newspapers and magazines and quickly became a national subject of news with outlandish advertising and publicity campaigns. The various electric companies cooperated in selling the idea of electric refrigeration, even if they competed on products. In 1928, a refrigerator was sent on a submarine voyage to the North Pole with Robert Ripley (of "Believe It or Not" fame).[22] "Selling refrigerators to Eskimos" became a common description of high-power sales techniques. Another promotion stunt was the formal presentation of a refrigerator to Henry Ford in a special radio broadcast of 1931.

The refrigeration industry was, of course, delighted with the way things were going, but the icemen were panicking. As electric refrigerators were refined and sales grew, the ice industry saw a threat to its traditional monopoly. Ice had once provided all the refrigeration used in homes; now, sales not just of ice but of iceboxes as well, plummeted. Until the early 1920s, icemen could laugh at the frequent failures of electric refrigeration in the home but by the late 1920s, it was no longer a joke. It seemed more like a serious threat. While some icebox manufacturers simply dropped the word "icebox" and began to refer to their products as "refrigerators," others fought back to defend ice. Household refrigeration pioneer Edmund Copeland described some of the obstacles that were thrown in his way by icemen:

> In many places the ice men were openly hostile to the new type of refrigeration. In certain cities, they put all kind of wiring obstacles in our way. They had ordinances passed to make our jobs difficult of installation. We were required to employ union men instead of our own to do the wiring. We had to employ licensed plumbers to install the machines instead of our own skilled and experienced men.[23]

The ice companies responded in their defense:

> A chunk of ice in a thoroughly insulated box remains, despite the wonders of electric refrigeration, the most efficient and the cheapest cooling method for the ordinary home. The iceman, individually, may resort to ridicule and vituperation; collectively, however, they are preparing to tell their story as never it has been told.[24]

The ice companies also fought back by providing better service and more efficient iceboxes. Both the ice companies and the electric refriger-

Ice ladies, ca. 1918.

ator companies published a number of advertisements that were derogatory to the other's products. One ice industry brochure warned that mechanical refrigerators caused cancer and requested that the U.S. Congress be petitioned to ban them.[25] Actually, the publicity being given to refrigeration in general, whether mechanical or natural ice, was increasing business for both. It educated the people in the need for refrigeration, and thousands who had never used ice, had no electric service, or who could not afford electric refrigeration, bought iceboxes and became customers of icemen[26] and women (yes, N.O.W.—there were some icewomen).

The public was well aware of the battle between the natural ice industry and the refrigeration industry. As late as 1932 it was still a matter of humor. In the Mack Sennett film comedy *The Dentist,* W.C. Fields played a dentist who becomes annoyed when his daughter wants to marry an ice man. She argues, "He goes to college. He's a Cornell man." Fields' character is not convinced. When she tells him her fiancé will have to return soon with a fifty-pound block of ice for the icebox, he retorts, "Keep that iceman out of here. I'm going to order a Frigidaire." Later, when his daughter elopes with the iceman, his daughter asks, "Father, you're not really going to buy a Frigidaire, are you?" Fields signals his approval of the marriage by answering, "Fifty pounds, and make it snappy."[27]

Little known to the natural ice industry at the time was another growing threat to them in addition to mechanical refrigeration: the early commercial use of dry ice. As mentioned earlier, French chemist Thilorier in 1835 had recorded the principle of solid carbon dioxide (CO_2)—dry ice. For the next 90 years it was observed in university laboratories but was not used for any practical applications, other than by a few doctors who used dry ice to remove warts. However, in 1897, a patent was granted in England to Herbert Samuel Elworthy, a doctor in the British Army Medical Corps, for the solidification of carbon dioxide to make soda water to mix with his whiskey. Thereafter, although the recreational use of whiskey continued to grow, few used dry ice for this patented purpose or any other.

The first commercial use of dry ice in the United States started in 1925 with Prest Air Devices, a company formed in Long Island City, New York, in 1923. The company made and marketed a successful fire extinguisher using compressed CO_2 and made solid dry ice for demonstration purposes. In 1924 it tried to sell dry ice to railroad companies to use for cooling refrigeration cars instead of regular ice, because dry ice had twice the cooling power of water ice, so it would be much more efficient. Based on this potential, Prest Air Devices was sold off and a new company was incorporated as the DryIce Corporation of America, which tried to patent dry ice again but was turned down by the Supreme Court. They did, however, trademark the name DryIce and constructed a dry ice plant in 1925 to produce it. The first test of dry ice in railroad cars was successful, and by 1932 there were 12 cars insulated properly to handle dry ice. But there was stiff competition. There were also 80 cars built with mechanical refrigeration, and 180,000 built to use regular ice, since the railroads invested in and owned many water ice plants on their property. Unfortunately for the new dry ice industry, mechanical refrigeration won out in the competition for the railroad industry.

But the DryIce Corporation of America had a plan B. They explored other potential clients. The first in 1926 was Schraff's stores, a confectionary store that sold, among other items, Eskimo Pie ice cream. Schraff's was experimenting with packaging to keep ice cream cold so that a customer could eat it at home instead of at the store. Before dry ice was used, salt was added to water and then frozen. This colder "brine ice" would keep ice cream frozen as it was distributed to retail stores, but when it melted it was corrosive, wet, and heavy. Dry ice was the ideal solution, and by 1927, many ice cream manufacturers, including Breyers, were using DryIce.

In 1929, DryIce teamed up with Liquid Carbonic to build 17 dry ice plants across the United States, including in Los Angeles, California. Birdseye Frozen Foods which started in 1931, immediately began using dry ice to transport its samples and promote its products. By 1932, with six or seven other manufacturers of dry ice, estimated U.S. production was more than 120 million pounds. Today there are thousands of dry ice producers around the world. Many sell to the public or industry, but some manufacture dry ice for their own use, such as in Disneyland.[28]

Meanwhile, while Einstein and Szilárd were developing their refrigerator patents in the late 1920s, other international inventors were developing alternative ways of refrigerating without electricity, since the rest of the world was far behind the United States in electrification. Two of these had similar designs: Australian Sir Edward Hallstrom developed his in 1923, and the other, David Forbes Keith of Toronto, Ontario, Canada, patented his in 1929. American inventor, industrialist, and entrepreneur Powel Crosley, Jr. (1886–1961), bought the rights to Keith's patent and manufactured the device in 1930, calling it the Crosley Icyball. He would produce thousands at two factories, one in the United States (called the Crosley Icyball device) and one in Canada (this device was called the Deforest Crosley Icyball).

Powell Crosley, Jr., had begun his career in 1907 by forming a company to build an automobile, which failed. After similar failed attempts he started a successful auto parts company. By 1919 he and his brother, Lewis, had sold more than a million dollars in parts and diversified into making phonograph cabinets. In 1921 he designed and manufactured radios and soon started an independent broadcasting station. By 1924, the Crosley Radio Corporation was the largest radio manufacturer in the world. In 1925 he introduced the Crosley "Pup" radio with a logo of a cute little dog named Bonzo with headphones listening to a "Pup" radio. Along with the Icyball in 1930, he began manufacturing full-size home refrigerators. He realized his youthful dream of building automobiles when he manufactured the small, inexpensive Crosley automobile from 1939 to 1952. He was the owner (1934–1961) of the Cincinnati Reds Major League Baseball team. Crosley Field, where the first night-baseball game was played in 1935, was renamed after him. His Florida winter retreat, Sea Gate, built in 1929 on Sarasota Bay for his wife, is a 2½-story Mediterranean Revival–style house with ten bedrooms and ten bathrooms. The State of Florida purchased it in 1991 and uses it as a meeting, conference, and event venue.

The Icyball consisted of two hollow metal balls, each approximately ten inches in diameter, connected to each other by a metal tube about one inch in diameter, separating the balls by about three inches. The connecting tube was bent into an inverted "U," rising vertically about three inches above both balls, with a wooden handle on top. The "cold ball" was a perfect, smooth sphere, with an opening in the bottom to connect to an ice cube attachment, and the "hot ball" was essentially spherical, but with vertical ribbing around its circumference to facilitate cooling. The entire assembly was about 18 inches high and 26 inches wide. The Icyball was an intermittent heat-absorption refrigerator, with a water/ammonia mixture used as a refrigerant. It was based on the same operational principle as Ferdinand Carré's machine of 1860, which was of a similar small size, but Carré used cylinders instead of balls. It was also suitable for household use and was marketed successfully for several years.

The Icyball operated similarly to the Carré machine. Both were portable refrigerators. The refrigerant of water/ammonia combined easily at room temperature in the "hot ball" but when the ball was heated for about 90 minutes (by a kerosene or gas burner underneath) the ammonia evaporated first because it has a lower boiling point than water. The ammonia vapor went through the tube to the "cold ball" and condensed, creating a cooling effect by removal of the heat, a process hastened by a cylinder filled with water inside the "cold ball." Within a few minutes the "cold ball" was cool enough for ice to form on its surface. When the balls were fully charged (indicated to the user by a whistle), the Icyball was lifted by its wooden handle and hung over the edge of an insulated box with the "cold ball" on the inside and the "hot ball" on the outside, with a hole in the lid of the box to allow the connecting tube through. As the ammonia vapor condensed, it returned to the "hot ball" to recombine with the water, and repeated the cycle. This process cooled the inside of the insulated box for over 24 hours, when the Icyball needed to be recharged again. Typically, it would be charged in the morning and

Icyball, ca. 1929 (courtesy Victoria Matranga and www.culinary.org at Kendall College; photograph Eric Futran).

would provide cooling throughout the day. An old Icyball found in the Toronto area was put through such a cycle in 1998 and produced a temperature of 18°F in the ice-cube tray hole with no insulated box to help the process. Quite a trick for a 70-year-old device![29]

As we saw, for the successful development of mechanical household refrigerators, engineers had to devise smaller alternative systems to the large refrigeration and ice machines, which were large, heavy, costly, and required constant service. The same thing was true for the development of air-conditioning equipment. A unit installed in the Minneapolis mansion of Charles Gates was approximately 7 feet high, 6 feet wide, 20 feet long, and possibly never used because no one ever lived in the house.[30] Early air-conditioning applications had numerous problems. Toxic refrigerants were hazards to health, automatic controls were needed to eliminate the operating engineers required for steam power, and the cost was prohibitive. As early as 1915, Willis Carrier had written: "If air conditioning is going to grow, there will have to be something done about mechanical refrigeration. We must have a refrigerating system that is simple and foolproof—so simple that we can run warm water through a pipe and have it come out cold." The refrigerating machine companies, satisfied with their low-temperature equipment, were not interested in developing new systems for such a small potential market as air conditioning. So Carrier decided that his firm had to design something that would meet all the requirements of air conditioning, reduce equipment cost, and improve reliability. He was aware of European work in centrifugal refrigeration and decided that this was the way to go.[31]

Centrifugal, or "turbo," compressors were used as early as 1890 for compressing air, but it was not until the early 1900s that they were used for refrigeration. The first device was probably by Maurice Leblanc of France, in the early 1900s, using water vapor and carbon tetrachloride as refrigerants. At about the same time, two Germans, Elger Elgenfeld and Hans Lorenz, independently experimented with centrifugal compressors as refrigeration devices. Lorenz published a paper in 1910 proposing the use of centrifugal compressors with various refrigerants. In 1908, W. Hessling received British Patent 13,109 for a multi-stage ammonia centrifugal refrigeration compressor. In Switzerland, Heinrich Zoelly designed a compressor before 1913 that established that a high-density refrigerant was required to keep the number of compressor stages to a minimum. But these European developments saw little use. Air conditioning in the United States was to be the first practical application of centrifugal compressors.[32]

A 1920 paper by Carrier revealed that he was ready to pursue design and manufacture of a refrigeration system using a centrifugal compressor. He could not find a suitable American manufacturer, so he selected the firm of C.H. Jaeger and Cie. of Leipzig, Germany, to manufacture a prototype machine using dichloroethylene, or "dielene," as a refrigerant. To the new compressor, Carrier added new heat-transfer devices, and the result in 1922 was a package chiller incorporating a centrifugal compressor, which had a 100-ton capacity and was covered by several patents. It went into production in 1924 when Carrier installed three centrifugal chillers in the J.L Hudson Department Store in Detroit, Michigan, with improvements of a spray-type cooling and dehumidifying system, which proved to be more reliable and less costly than any on the market.[33]

But the system was still far too large and unsuitable to be used in private homes, and as the public became aware of air conditioning in their local movie theaters, their desire for cooling their homes created a potentially huge market. Such cooling was a technical impossibility until the late 1920s, when innovative developments in household refrigeration could be applied to room cooling with devices that were compact, reliable, and affordable. Still, the potential market was there. In 1918, inventor

Carrier centrifugal compressor chiller, 1922.

Alexander Graham Bell had told a graduating class of students: "The problem of cooling houses is one that I would recommend to your notice, not only on account of your own comfort, but on account of public health as well."[34] This was during the influenza epidemic of 1918, when people became highly conscious of personal cleanliness and sanitation in the home to avoid germs. Household cooling was therefore seen as a means not only for reducing heat and humidity but also for reducing the side effects of improper ventilation, which, according to *Scientific American*, became known as "elephant air": "Elephant air derives its name from an anecdote about the small boy who was invited to a birthday party a few weeks after having seen his first circus. Upon entering the home to which he had been invited, he shouted with glee, 'Oh! They're having a circus. I can smell the elephants.'"[35]

The first patent for a through-the-wall room cooler was by Gustave Kramer in 1918 (U.S. Patent 1,706,852, *Room Cooling Apparatus*), but the technology to produce it was not available at the time until 1928, when Frigidaire, the dominant refrigerator manufacturer, would acquired it. Frigidaire produced a "split-system room cooler" in 1929. The air conditioner weighed 200 pounds and was about the size of a refrigerator. It used sulfur dioxide refrigerant and had a capacity of one ton. It was rather expensive and required a complicated installation. Another refrigerator manufacturer, General Electric, also became interested in room coolers at about the same time. In 1928 GE development engineer Frank Faust was asked "to develop a demonstration of the feasibility of applying the GE refrigerator development to room cooling." Faust designed an experimental water-chiller using the Monitor Top household refrigeration unit, the chilled water being

Frigidaire room cooler, 1929.

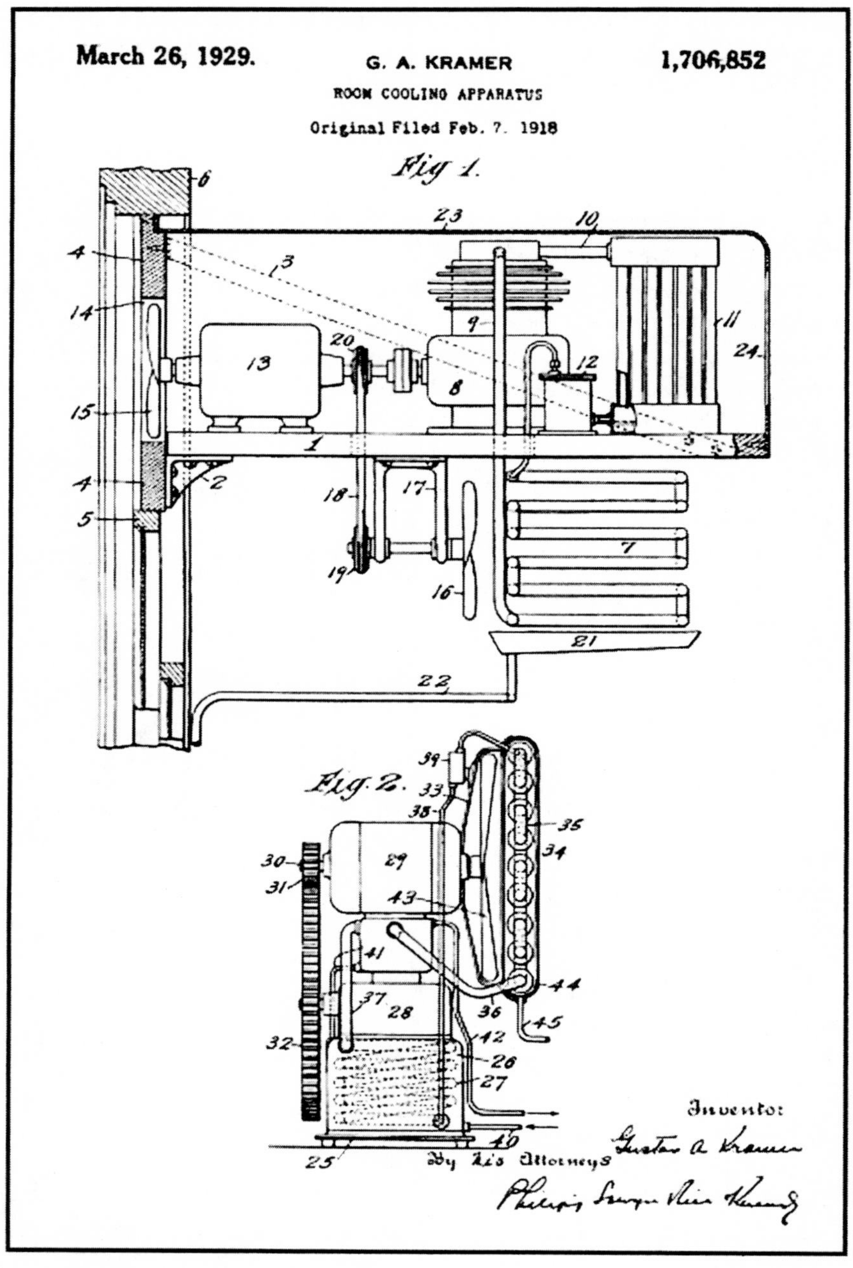

Kramer room cooler, U.S. Patent No. 1,706,852, 1918.

pumped to a finned coil in an air handler that had been obtained on consignment from Willis Carrier. The system was installed in the office of a New York psychiatrist friend of GE president Gerard Swope. Faust recalled:

> I personally tested the system and prepared a technical report. The system worked perfectly. The demonstration apparently was very successful in convincing the top officials of the company to finance further development of room cooling. It may also have influenced Carrier to develop their so-called "Atmospheric Cabinet," first installed in 1932.[36]

Faust then was asked to design a room cooler that used the Monitor Top sulfur dioxide compressor, a water-cooled condenser, and a thermostatic expansion valve with a finned-coil evaporator and fan. In 1930 and 1931, 32 prototypes were built in specially designed wooden housings resembling radio cabinets. On this basis, GE established its air conditioning department in 1932.

GE room cooler, 1932 (courtesy the Museum of Innovation and Science, formerly the Schenectady Museum, Schenectady, New York).

Meanwhile, Carrier and his staff had been so busy with industrial and movie house air conditioning that they were slow to enter the residential and light industrial room-cooling market. Their first venture was the Unit Air Conditioner, a spray type, designed by Carlyle M. Ashley (U.S. Patent 1,883,456), which was introduced in 1928 for offices and factories. Called the Centijector within the company, it used a finned coil for heating. A water spray provided humidity control in winter and, using chilled water, could lower both humidity and temperature in the summer. By 1932, Carrier would introduce the Atmospheric Cabinet or Room Weathermaker, which had a cabinet design similar to those being sold by Frigidaire and General Electric. Carrier recognized the potential of small air-conditioners by hiring "America's first woman air conditioning engineer," Margaret Ingels, to head a campaign to educate the public to the benefits of air conditioning.[37] She had previously been a staff engineer at the American Society of Heating and Ventilating Engineers from 1921 to 1927.[38]

Women would become very important to the sale of household products. The commercial success of mechanical products in general throughout the 1920s was enabled by reasonably good economic times, and in particular by the willingness of people to spend money on consumer products such as automobiles, toasters, washing machines, waffle irons, clothes irons, fans, power tools, vacuum cleaners, and, of course, refrigerators. It was not called the Machine Age for nothing. All these products were popular because they saved manual labor or enhanced daily life for both men and women, but because most of these new products were household appliances used mostly by women, it was women who seemed to benefit the most. One 1929 study reported that most families of students at Mount Holyoke College in Massachusetts (a women's college founded in 1837 that attracts mostly upper middle class students) had acquired four or five major appliances between 1919 and 1929.

Women welcomed the introduction of mechanical and electrical household appliances for their obvious advantage of reducing tedious and ornery traditional household tasks. But they were also influenced by the growing home economics movement, started in 1908 by Massachusetts Institute of Technology chemist Ellen Swallow Richards (1842–1911), as the first president of the American Home Economics Association. The organization was founded to improve the conditions of housework through research, education, and the application of scientific principles to nutrition, sanitation, and physical fitness. The organization inspired the establishment of home economics classes in thousands of high schools and colleges. By 1920, many of these students were busy starting their own households and enthusiastically adapting to modern conveniences available on the market.

By 1920, U.S. industry was adopting the principles of Taylorism, the scientific process of making production processes more efficient, pioneered in 1911 by Frederick Winslow Taylor (1856–1915), who inspired Henry Ford's assembly line innovation for his 1913 Model T; this set the precedent for mass production, which became known around the world as Fordism. Women followers of Taylor such as Frank Gilbreth's (1868–1924) wife, Lillian (1878–1972), focused on kitchen and household organization to improve efficiency. Another follower of Taylor was Christine Frederick (1883–1970), household editor of the *Ladies' Home Journal*, who had been writing articles that applied scientific principles to household management. By 1920, Frederick had published an influential book, *Household Engineering: Scientific Management in the Home*. She advised women how to arrange kitchen equipment for efficient traffic patterns, identified laborsaving devices and procedures for use, and urged women to develop skills of analysis, study, and planning to better manage their household tasks, including the efficient and detailed interrelated steps of food preparation (4 sub-steps) and clearing away (4 sub-steps) afterwards.

Frederick was also following in the footsteps of home economist Catherine Beecher, mentioned earlier, who had with her 1869 book *The American Woman's Home* pioneered techniques in the organization of the kitchen and home for more effective and efficient operation of household tasks by women. Frederick improved efficiency on all these tasks but also identified the many shortcomings of functional household products that were a result of neglect by manufacturing engineers, who paid no attention to user needs, largely because they were men obsessed with mechanical efficiency alone:

> Very often a device which fulfils other conditions ... fails in the small but essential point of comfort in use. This is especially true of handles, levers, etc., which either by their shape, finish, or point of attachment prove uncomfortable when used in the hands of the worker. There is the case of a bread-maker with the leverage applied at the top of the pail; otherwise a fine labor-saver, it requires an awkward arm motion which would not be the case if the leverage were applied at the base and side of the pail as indeed it is, in another make. The handles of many eggbeaters, mashers, spoons, etc., are not shaped for the comfort of the hand, although there are others on the market, which do offer this point of comfort. Sometimes the handle is too short or too long, flat instead of rounded. Or a lever would be easier to operate if several inches longer, and many other instances occur where the small but important points of comfort are not considered.[39]

Frederick wrestled with the problem of the height of standing worktops in the kitchen, which ranged from 31½ inches to 34¼ inches off the floor. She concluded that "no absolute rule can be given invariable heights because not only the height of the worker must be taken into consideration but also the length of her arm, and whether she is short or long-waisted, etc."[40]

This all sounds a bit like an article in *Consumer Reports*, a well-known magazine published by the Consumers Union, which still evaluates competitive consumer products; but that organization and magazine would not be founded until 1936. Frederick was anticipating a practice that also would not become commonplace until the late 1920s, when manufacturers engaged designers who not only would consider the overall style of products, but would also address user aspects such as Frederick suggested. These matters would then become known as "human factors" and later as "ergonomics." Frederick educated women to look critically at their new electrical and mechanical products not just as laborsaving devices, but as modern technology that needed to consider women as primary users and purchasers.

Frederick's writings, part of a trend that called for viewing housework as a profession, were translated into German and were instrumental in the development of the Frankfurt Kitchen in 1926 by Austrian architect Margarete Schütte-Lihotzky (1897–2000). Her design was for large-scale project apartments in Frankfurt, and thus had to be compact, efficient, and versatile. She did detailed time-and-motion studies to determine how long each food-processing step took, and planned the design to make it easier and faster for women to prepare meals. There were dedicated, handy storage bins for flour, sugar, rice, etc., to keep the kitchen tidy, and the kitchen, being quite small, required much less time and walking effort. The Frankfurt Kitchen was installed in some 10,000 units in Frankfurt and was a huge success. Kitchens were being transformed from traditional open-hearth fireplace rooms, family centers where housewives spent most of the day preparing food, to efficient machines for cooking that required less time and effort for working women. By the mid–1930s, women's influence on household mass-producing manufacturers of appliances would become overwhelming.

The laborsaving qualities of appliances appealed mostly to those in the upper middle class who were no longer able to afford servants.[41] The use of domestic servants was one of the social issues that changed significantly in the 1920s. In the earlier years of the century, "decent" housewives were never depicted as doing the heavy work of the household themselves because it was assumed that mostly immigrant household servants would do so. When women's magazines offered instructions for proper laundry work or cleaning, phrases such as "instruct your laundress..." or "see that your maid..." were often added.[42]

But attitudes were changing with the reality that the costs of servants were increasing as their availability diminished, first by World War I and then by immigrant restrictions in the mid–1920s that drastically reduced the influx of foreign-born young women. From 1920 to 1924, wages for servants had increased five-fold, compared to the mere doubling of middle-income wages during the same period. The Mount Holyoke study of 1929, mentioned earlier, also reported that 96 percent of the studied families had decreased the amount of household help. Naturally, the electric utilities and electric manufacturing companies recognized this by advertising their products as "servants in the home."

Frigidaire combination refrigerator/stove, 1929.

The supply of servants was further reduced by the expanding opportunities for unskilled women in factories. Only one quarter of the 700 urban households of college-educated women that were studied by the U.S. Department of Agriculture in 1930 employed a domestic servant. This situation would only get worse in the 1930s.[43]

Economic factors were at work, as well, during the 1920s. Due to the booming economy, incomes were relatively steady and rising. Most people could find work if they wanted it, and the desire for the new machines, particularly the automobile, but new household appliances as well, became a powerful incentive to become two-income families, or to seek higher-paying jobs. Possessing modern conveniences was the way to publicly demonstrate upward mobility into or higher up in the growing middle class. Indeed, refrigerators were replacing automobiles as the latest "must have" modern technological marvel. A contemporary publication illustrated how a refrigerator was considered to be a step up in social status:

> "My dear, you must come over and see our new artificial ice and electric stove combination with radio attachment that broadcasts as ice freezes! Social life in the suburbs has become awfully hectic since anybody who is anybody has gone in for plain and fancy ice plants. Time was when you could get a line on a family's social status by the make of their car.... Nowadays you can tell whether the so and so's are worth knowing by the number of ice cubes per minute their ice box is throwing off."[44]

In fact, combination refrigerator-stove appliances were introduced by Frigidaire and General Electric in about 1929, but were never popular. The next chapter shows how the arrival of fashion in the 1920s elevated women to the level of primary consumers of household products. The public demanded new designs, and manufacturers needed to turn to a new breed of designers to sell their products in the midst of the Depression.

Chapter 6

The Rise of Fashion and the Depression

The Roaring Twenties not only meant economic success and the absence of war, but the roar of Tommy guns spawned by Prohibition, which was enacted by the government in 1920 and criminalized the public serving of alcohol. This led to moonshine stills in the countryside, illegal bootleggers, and warring gangsters such as Al Capone and Bugs Moran committing theft and murder to monopolize the liquor business in major cities like Chicago. On the other hand, the 1920s also included unforgettable and dramatic historical moments. In 1927, a banner year, Charles Lindbergh thrilled the nation with his solo flight across the Atlantic, the forerunner to public air travel; the New York Yankees won 110 games and the World Series with superstars Babe Ruth and Lou Gehrig; movies introduced sound with the Al Jolson film *The Jazz Singer*; and Philo Farnsworth (1906–1971) invented electronic television, although most people at the time had no idea what television actually was, let alone what it could become. Radio, since its introduction in 1920, was the primary popular public communications media.

During the 1920s, there was a significant increase in public awareness of cleanliness and sanitation. After the influenza epidemic of 1918, people had become highly aware of the importance of personal cleanliness and sanitation in the home to avoid germs. Dixie Cups made of paper (invented and produced first in 1908) reduced the previously common practice of drinking from community tin cups at public water sources, and by 1920, the public drinking glass disappeared from railway carriages, and the tin cup from the backyard farm pump. Bathroom fixtures such as sinks, toilets, and bathtubs were finished in white, sanitary, porcelain enamel for easier cleaning and a sanitary appearance; kitchen sinks, likewise. White, of course, was the traditional color of sheets, hospital beds, laboratories, and nurse's uniforms, all suggesting cleanliness and technology.

Right after World War I, 80 percent to 90 percent of all porcelain enamelware was in pots and pans. The work surface on old kitchen cabinets originally was a metal finish called Nickeloid. But new technologies had expanded the production of porcelain-enameled coatings on large surfaces such as tabletops, signs, architectural panels, etc., to a degree not previously possible. Around 1920, this led to the creation of a porcelain-enameled-top kitchen table, the most popular size of which was 25 by 40 inches.[1] The color was, of course, white.

This trend came into play in the appliance industry, where major appliances are also still known today as "white goods." As we know, ice-boxes, and even mechanical refrigerators, until the 1920s, were almost all made of wood (imitating furniture, which is still called "brown goods"). Other major metal appliances, such as electric stoves, were also in black or brown until 1922, when the Edison Electric Company, manufacturer of Hotpoint stoves, received a custom order from a utility executive for a white porcelain stove. Rather than admit that such a stove was not being produced, Edison quoted an exorbitant price to discourage him. Upon receiving an order despite the high price, Edison had to develop a new annealing process to apply white porcelain enamel to sheet metal. The next year, Hotpoint models included white ranges with nickel trim on the doors, and by 1925 white became a stove industry standard.[2] So when refrigerators began to be produced in high volume, as typified by the Monitor Top in 1927, only steel cabinets, easily formed and shaped, made this economically feasible. White quickly became the standard major appliance color, including for refrigerators; porcelain enamel became the standard surface finish, following the trend of bathroom fixtures; and the public perception that white signified cleanliness became indelibly reinforced.

But no sooner had white become standard than a trend began toward color in August 1927, soon becoming what some called a color craze. The *New York Times* wrote:

> One may have, of course, the kitchen which suggests obviously that it is the food laboratory of the home ... in glistening white ... yet as our bathrooms are changing in spirit from the old, ultra-chaste whiteness to more gracious and gayer interiors, so our kitchens are adding to their efficient equipment the charm of color.[3]

Apparently, some women were becoming fed up with what they called "poorhouse white" in their kitchens; it was too much like a laboratory. The October 1927 *House Furnishing Review* reported: "There is no mistaking the importance of the color trend that is sweeping over New York. While every large metropolitan department has gone in for color to an impressive degree, Macy's and Wanamaker's are outstanding examples."[4]

This trend, however, did not affect major appliances such as refrigerators, which remained in conservative white. It affected primarily bath and kitchen accessories, such as small appliances. Unfortunately, merchandizing chaos soon erupted. Stocking a wide range of colors in merchandise caused huge inventory problems, and worse, there was an increasingly widespread variation of the same color from different manufacturers. And none of them matched. Stores resorted to having store painters repaint items in order to have them match and balance out stock. The problem was that there were no official or accepted color standards. The situation for the housware industry worsened for a decade until 1937, when Eliot Walter from Macy's was named chairman of a Color Standardization Committee, which, working within the U.S. Commerce Department, established kitchen and bathroom standards: kitchen standards included White, Kitchen Green, Ivory, Delphinium Blue, Royal Blue, and Red. The Commerce Department published two booklets, "CS62–38 Colors for Kitchens Accessories," and "CS63–38 Colors for Bathroom Accessories."[5] These were essentially forgotten during World War II, when manufacturing nearly stopped, and in 1946, *House and Garden* magazine would have to start again from scratch. This time around, major appliances, including refrigerators, would be impacted.

Color, as a decorative element, was typically the prerogative of housewives, which suggested that women consumers were the source of the trend toward color. Perhaps more significantly, the role of women was changing dramatically in the 1920s, partly as a result of women's being granted the right to vote at the start of the decade. Women felt more empowered to participate in family financial decisions and to express their own independence. The younger women felt free to celebrate this independence by wearing hairstyles and clothing that were shocking to many. These young ladies were called "flappers," a slang term that emerged in England in the 1890s and was used for young prostitutes or any lively teenage girl. The term was popularized in America by the 1920 Francis Marion film *The Flappers*.

American flappers were imitating fashions initiated in France, the center of style, by fashion leaders such as Gabrielle "Coco" Chanel (1883–1971) and African American jazz entertainer Josephine Baker (1906–1975) at the Folies Bergère in Paris. Baker popularized an American dance step called the Charleston that appeared in the 1923 Broadway show *Runnin' Wild*. American jazz was the most popular music of the time (the reason the 1920s were also called the Jazz Age), but the slang term "jazz" also meant anything exciting and fun. And young women expressed this by defying traditional styles and acceptable behavior. They wore short skirts and cloche (helmet-type) hats, they bobbed (shortened) their hair, and they listened to jazz incessantly. They wore excessive makeup, drank liquor, smoked, drove automobiles, and treated sex in a casual manner. Their behavior was perceived by many to be outlandish, sexually provocative, frivolous, and irresponsible, much in the same way as "hippies" would be viewed by their conservative elders in the 1960s.

But flappers were merely expressing their freedom from traditional Victorian restraints and eventually would convince their elders to loosen up a bit. More significantly, they would establish fashion as a driving social force of the 20th century; it would soon have an enormous financial impact on all the manufacturers of consumer products, including refrigerators. This influential transformation began, oddly enough, in 1925 with the Exposition International des Arts Décoratifs et Industriels Modernes (International Exposition of Modern Decorative and Industrial Arts) held in Paris that year. The lengthy title would soon be shortened to Art Deco Exhibition. France wanted to promote and reestablish its prewar reputation as the global leader in style, fashion, the decorative arts, and the fine arts.

After all, French artists had pioneered in Impressionism in the 19th century, the Art Nouveau movement at the turn of the century, and Modern Art in 1913, with its abstract paintings defying tradition and public sensitivities. When European "modern art" was first shown in New York at the Armory Show in 1913, President Theodore Roosevelt represented most Americans when he said after viewing the exhibition: "That's not Art!" Most Americans considered traditional landscapes and work by the great masters as art. Many serious American artists were, of course, aware of French leadership in the modern arts and flocked to Paris in the 1920s to study art and absorb the artistic and exciting culture. But Americans in general were blissfully unaware of the latest abstract modern art movements such as Fauvism, Expressionism, Futurism, Surrealism, and Cubism, which disassembled a traditional image into overlapping planes with both front and side views of a face superimposed.

Unlike previous world fairs, which focused on technical accomplishments and traditional art by various countries, Art Deco organizers wanted to establish modern decorative design standards. Exhibitors were required to display *only* "those exhibits that were not dependent of the art of the past." Twenty-four countries accepted invitations, but China and the United States declined. China's excuse was political disruptions, but the United States had to admit that its manufacturers simply had no commercial interest in attending (they were doing just fine with business at home), and besides, had no new innovative designs to display. Actually, industry had many new mechanical, scientific, and functional designs to display, but in the context of the art world and Art Deco, "designs" meant artistic designs: the way things looked, not what they did. And in that category, U.S. manufacturers were inclined to simply copy previous artistic European designs, not invent new ones.

Some U.S. art educators and the cultural establishment realized that the United States was far behind Europe in both fine art and the design of decorative consumer products, and when they asked Herbert Hoover, then secretary of commerce, in charge of the Coolidge administration's industrial policy, to delegate a commission of 100 representatives of trade associations, along with some artists and designers, to visit and report on the Paris exposition, Hoover gracefully complied.

The exposition ran for six months, May through October, attracted 16 million visitors, and displayed the work of many French and European glassware, jewelry, furnishings, and interior designers. Famous Paris department stores such as Bon Marché featured exhibits of household goods from the exhibition in "ensembles," that is, complete rooms furnished in the latest fashion with modern accessories and furniture. Many of the designs were by acclaimed designers and architects who worked in leading European design schools such as the Bauhaus in Germany, or the Vkhutemas in the new Soviet Union. Other independent designers included French glass designer René Lalique (1860–1945), Danish designer Kaare Klint (1888–1954), Italian designer Giovanni "Gio" Ponti (1891–1979), and the famous French modern architect, who used the pseudonym, Le Corbusier (1887–1965), although his given name was Charles Édouard Jeanneret Gris. Corbusier designed an influential exhibit, Pavilion de L'Esprit Nouveau (Pavilion of the New Spirit), in which he furnished an urban apartment as a "machine for living." The plain, unfinished white-wall interior was furnished with simple commercial products such as metal filing cabinets, hospital tables, and laboratory flasks as vases. It was a complete rejection of traditional decorative arts, but was more influential to the modern movement than anything else in the exposition.

Corbusier, like a number of other European architects, was fascinated with the visual language and social impact of American industrialization and mass production as it had evolved in the early 20th century. To many European intellectuals, images of towering Western grain elevators, massive factories for mass production in Detroit, streamlined aircraft, and racing cars were all representative of 20th century modernity and progress for mankind. German architect Walter Gropius (1883–1969) was similarly inspired by American industrialization when he initiated the Bauhaus (Building House) in Weimar, Germany, in 1919 after World War I as an architectural, art, and design school. Germans believed they could best restore their prewar industrial leadership in the world with the American system of manufacture based on Fordism and Taylorism. Gropius referred to America as "the motherland of industry." He sought, through his design program, to achieve the nationalistic goal articulated by his prewar educational predecessor, Herman Muthesius (1861–1927), who wanted to unite art and industry for national economic advantage. As early as 1903 he called for "a new style, a *machinenstil*" (machine style) whose shapes were "completely dictated by the purposes they serve." Gropius would by 1924 have succeeded in demonstrating to the architectural world that simple, geometric forms were indeed a visual expression of pure functional purpose and industrialization.

After the 1925 exhibition, the American Commission in its report warned that the modern movement in applied arts would soon reach the U.S., and that

> [Americans needed to] initiate a parallel effort of our own upon lines calculated to appeal to the American consumer.... As a nation we now live artistically on warmed-over dishes. In a number of lines of manufacture we are little more than producing antiquarians. We copy, modify and adapt the older styles with few suggestions of a new idea.... It would seem ... that the richness and complexity of American life call for excursions into new fields that may yield not only innovations but examples well suited to the living conditions of our times.[6]

The few American designers of commercial products such as lamps, furniture, glass and ceramics who visited the Art Deco exposition returned committed to the new modernist styles of applied arts. "Applied art" was the term for what artists who worked in industry did— applying their artistic talent to make products look more attractive, primarily through two-dimensional surface decoration. What was it about modern design that so impressed both visitors and critics? Art critic Helen Read, who reported on the exhibition, tried to explain the new forms of the "modern movement" in Paris:

> It is characteristic of the new décor that the nature of the material is at all times respected and is allowed to dictate its treatment. Wood is wood, iron is iron; the bad taste that invariably ensues when the attempt is made

> to give the quality of one material to another, to make wood look [like] iron or marble like lace, is avoided. This universal tendency toward simplicity of outline should be of the greatest interest to the American designer if he will recognize that this is the fundamental note. All design in this country is governed by the factor of whether or not it can be reproduced in mass production. It costs no more to get out a good design than it does a bad one, and the fact that the best designs of the new décor are the simplest to the point of being geometric, makes them so much the more easy to put on the market.[7]

It was not long before these eclectic new styles were generally referred to generally as Art Deco styles. Some resembled the simple, abstract geometric forms that characterized the recent European fine art painting styles, such as Cubism. Others looked like the rectangular abstractions of the Dutch de Stijl paintings. Product forms were simple, plain, and devoid of traditional surface decoration of patterns, flowers, or ornament. Art Deco was more about product form and character than about surface or cosmetic decoration. By January 1926, a *New York Times* editorial about the Paris exposition urged industry to change because the public would soon catch up with Europe in recognizing the competitive value of new designs:

> American captains of industry must soon change their traditional characters. True, the public may still cling to its conception of them as iron-jawed, level headed giants of efficiency ... but if the present international competition keeps up, we may find them, like the Germans, openly cooperating with museums and art schools; or, like the French, proclaiming themselves merchants of beauty.[8]

Advertising agencies quickly got the message. Ernest Elmo Calkins (1868–1964), head of the Calkins and Holden agency, and regarded by many as the "Dean of Advertising Men," immediately established an internal product design group within his agency headed by magazine editor Egmont Arens (1889–1966), and in the August 1927 issue of the *Atlantic Monthly*, Calkins wrote an influential article, "Beauty, the New Business Tool," in which he stated the case for more attractive products:

> We passed from the hand to the machine, we enjoyed our era of triumph of the machine, we acquired wealth, and with wealth, education, travel, sophistication, a sense of beauty; and then we began to miss something in our cheap but ugly products. Efficiency was not enough. The machine did not satisfy the soul. Man could not live by bread alone. And thus it came about that beauty, or what one conceived as beauty, became a factor in the production and marketing of goods.[9]

Calkins continued in the article:

> The only art that can survive and grow is art that is related to our life and our needs, and that has a sound economic foundation. It is to be hoped that manufacturers in the search for design to beautify their products will start with a clear conception of what beauty is, especially beauty in an article of use ... the surest guide in divining new beauty in machine-made things is to grasp and interpret the beauty they naturally and intrinsically possess.[10]

In other words, beauty in useful (not decorative, but functional) products is not in its extraneous artistic decoration or embellishment, but rather is in the honest and simple visual expression of its natural function through form and character. Whether Calkins knew of the German Bauhaus or not, he was expressing exactly the same philosophy of design, but in business terms. Calkins asked his advertising colleagues: "How could [modern] ads work, if the styling of products did not express the qualities claimed for it?"[11]

U.S. department stores and museums soon began to exhibit designs and collections from the exhibition, promoting them as the latest styles from Paris. Typical was the May 1927 Macy's Art and Trade exhibition in New York, which was heavily promoted by radio interviews with participants and which drew 40,000 people. Similar promotions of the new designs occurred at other major department stores such as Franklin-Simon in New York, Wanamakers in Philadelphia, and Marshall Field in Chicago. In 1928 the B. Altman department store in New York held an exhibition called Twentieth Century Taste. The promotion of new Art Deco styles was not the only focus of exhibitions. One tied the modern age to the visual and artistic character of functional machines. Jane Heap, editor of a literary journal, *The Little Review*, organized a Machine-Age Exposition in May 1927 in Steinway Hall, a commercial building in midtown Manhattan. Influenced by Le Corbusier's 1925 "machine for living" at the Paris Exposition, Heap displayed actual machine components, such as a Studebaker crankshaft, a Curtis aircraft engine, and a Hyde Windlass propeller, along with photos of skyscrapers, and the Bauhaus in Dessau. Suddenly, the simple, functional look of machines became an expression of modern art, as well. This concept of functional simplicity and beauty would soon dominate the world of manufacturing, thanks to a few American designers who recognized that ordinary, functional, mechanical devices could not only become works of art, but also generate impressive sales success.

In 1928 Franklin-Simon department store in New York engaged famous stage designer Norman Bel Geddes (1893–1958) to design a modern approach for window displays, featuring dramatically lit, very modern-looking, cubistic pyramids and metallic mannequins. When they were revealed to the public, they caused massive traffic jams and crowded sidewalks. Within 90 days, every store on 5th Avenue had changed its windows to the new modern style. That same year, Macy's staged the International Exposition of Art in Industry, exhibiting modern décor by famous European architects and designers. The Metropolitan Museum of Art in 1928 staged an exhibition called Contemporary American Design, emulating the 1925 Paris exposition with French room ensembles. The Museum of Modern Art opened in 1929 to recognize and celebrate the latest trends in art and design, and specifically promoted modern design.

Consumers responded positively to the extensive promotion of new modern styles, particularly fashion conscious women such as flappers. In addition, movie and theater sets began to feature interior decoration, furniture, and home accessories that were in the modern styles, which were viewed and admired by millions of moviegoers. People wanted what they saw. The styles spread to various manufacturers, anxious to keep up with new styles. U.S. designers of decorative arts worked for the "art industries," specifically those that made decorative home products, such as lighting fixtures, furniture, art metalwork, carpets, drapery, wallpaper, textiles, silverware, ceramics and jewelry. Understandably, many of these designers were women, because women, who traditionally were responsible for maintaining the home, often purchased these kinds of products. Always looking for the latest style trends, these designers quickly began to adapt their designs to Art Deco. This was the practice for generations of the "artists in industry" working in the industrial art field, who were professionally trained in art and design schools. As the modern styles became more and more popular, consumers responded with enthusiasm and the result was higher sales.

In the late 19th century, when industry began to produce highly mechanical and electrical consumer products, they had been designed almost exclusively by engineers, without any artistic considerations, because they involved scientific, technical, mechanical or electrical components. Engineers were almost exclusively men, highly technical men, whose sole purpose was to make things work effectively and to be produced economically by mass production. Appearance, style, and convenience to users were totally ignored, because the products were primarily functional, not decorative. These were products such as locomotives, automobiles, stoves, cameras, telephones, sinks, toilets, office machines, typewriters, and, of course, refrigerators. Their primary appeal in the market was their functionality, not their appearance. So early on, these manufacturers were arbitrarily placed into a new category called the "artless industries" to distinguish them from the traditional "art industries."

Understandingly, these "artless industries" were the bottom of the food chain for artists, who generally shunned them because of the dirty, noisy, and engineer-run factories they would have to work in. While some decorative artists happily worked for the art industries—many of which, by the way, also required working in factories—most artists preferred the fine arts, where they could have a quiet, clean private studio to work in, could exhibit their work in galleries, and could acquire a personal and professional reputation. So the "artless" industries had absolutely no artists or industrial artists either on staff or as consultants, nor were there any artists interested in working there. Generally, the functional products of these industries were doing quite well in the 1920s, because more and consumers were buying as many of these new marvels of technology as manufacturers could produce. The reason was that most "artless" industries, through the economies of mass production, were able to offer products to consumers that were previously affordable only for the well-to-do and were luxuries for the middle class. People bought them because they could afford them. Most manufacturers had absolutely no incentive to make any appearance changes in their designs. Why change designs with expensive new tooling costs when they were selling at record levels as they were?

The first "artless industry" that recognized the importance of style was the automotive industry. After World War I, automobile architects were hired to design high-price custom car bodies for Hollywood celebrities, in the same way that architects designed buildings: with pre-planned drawings and featuring much more attention to exterior styling. The car manufacturers then mounted the custom bodies on production mechanical frames. In 1925, General Motors (GM) head Alfred P. Sloan decided to use the same techniques on production cars, as he implemented the first annual model changes in the industry. The first was the 1926 Chevrolet, which differed noticeably in style and color from Ford's Model T, which had not changed significantly in appearance since 1913. In 1926 Chevy sales doubled and Ford sales dropped by 25 percent. Convinced that styling was the key to increased sales, Sloan hired Hollywood custom car designer Harley Earl (1893–1969) in 1927 to initiate a new Art and Colour Section at GM to design all new models. Earl first designed the sleek 1927 LaSalle, which was an instant success. By 1928, Earl had a staff of 50 people and would dominate car styling for decades. GM's success led other "artless" industries to follow its example and look for artistic help.

A few enterprising designers, mostly in New York City, who had seen the oncoming trend of modern design in the "artless" industries, decided to abandon their normal art-related professions and establish consulting offices specializing in design for the "artless" industries, where the need for appearance design was greatest. They were willing to go into factories and lock horns with the engineers, many of who resented artists' trying to tell them how to design products. Norman Bel Geddes (1893–1958), the preeminent stage designer, and French-born fashion illustrator Raymond Loewy (1893–1986) both entered the automotive design field as independent consultants (Bel Geddes in 1927, Loewy in 1929), but soon were designing furniture and office equipment as well. Advertising artist Walter Dorwin Teague (1883–1960) landed a contract with Kodak designing Art Deco cameras in 1927. Henry Dreyfuss (1904–1972), also a successful stage set designer, won a contest in 1929 for designing the "phone of the future" for Bell Labs and opened a consulting office.

All these new pioneer designers adopted the relatively obscure term of "industrial design" for what was essentially a brand new specialized art profession servicing the artless industries by modernizing and improving the appearance of their mass-produced, mechanical product designs. Soon there were many of these industrial designers coming from a variety of traditional artistic fields of design. Industrial designers not only modernized and simplified the external style of the product, but also addressed the type of user needs or "human factors" as suggested by Christine Frederick by making the product easier to use, more understandable, more convenient, and more comfortable for the user, and often reduced the manufacturing cost by eliminating or combining components.

The public demand for modern design expanded rapidly from 1927 to 1929. Consumers, becoming accustomed to modern design through department store promotions, films, advertisements and purchases, began to realize that their latest electrical or mechanical products looked terribly mechanical, awkward, clunky, and rather ugly, compared to their new, sleek, modern furniture and household décor. They started to demand more modern styles in their "artless industries" automobiles, vacuum cleaners, and kitchen-tools such as mixers, toasters, griddles, telephones, clothes washers, stoves and refrigerators.

On the other hand, many manufacturers were like a deer caught in headlights when they became aware of this new public demand for style. They were accustomed to consumers who were happy to get any product that functioned well. Most engineers did not have a clue as to how to improve the styling of their products. They were totally unaware of industrial designers. They had no artistic people on staff. The only drawings they knew how to make were mechanical drawings of functional components. Although many were reluctant to change because of the cost of new tooling, some decided to do so, and those that did saw their sales increase dramatically, from 25 percent to 200 percent. This convinced a few other manufacturers to try industrial design, but for most, it was business as usual.

The early demonstration of dramatic sales increases certainly benefited the very few practicing industrial designers, whose fees soon became astronomical not only compared to ordinary, decorative art-industry designers, but to corporate executives, as well. This was in a time when corporate executives were making $5,000 a year. Average annual retainer fees of industrial designers were $50,000 and hourly rates were $50. In today's dollars, these amounts would be $687,000 and $687, respectively. These lucrative salaries and consulting fees attracted more designers into the field because of the increasing demand for industrial designers in industry.

Still there were only a dozen or so practicing industrial designers, who were immediately swamped and had to turn away many clients. Manufacturers sought out whosoever they could find. The American Radiator and Standard Sanitary Corporation, which made bathroom fixtures, hired a decorative glass designer, George Sakier (1897–1988) to design their products. The American Management Association (AMA) began to promote modern design with seminar titles such as: "How the Manufacturer Copes with Fashion, Style, and Art," "The Renaissance of Art in American Business," and whether "intelligent and profitable use can be made of the modernistic style of design." At the AMA annual conference on October 29, 1929, members cited examples of increased sales in specific industries simply because of the use of industrial designers. The keynote speaker, an advisor on merchandising problems for the Associated Industries of Massachusetts, stated that modernistic style was the "offspring" of the Jazz Age and that "modern style will remain as long as the machine in its present form is the characteristic of world production, [changing] its outward form from year to year, just as women's dress styles change."[12]

By one of those strange coincidences of fate, just as modern design was becoming influential in industry, the date of that AMA conference was the exact same date, October 29, that the stock market crashed, a day later known as "Black Tuesday." The stock market crash itself was caused by permitting investors to buy stock on margin (credit). Many brokerage firms were lending small investors more than two-thirds of the face value they were buying. Thus, many investors had much more than their actual cash value invested in the market. This was a fantastic deal as long as the market was rising throughout the 1920s. During that time, the Dow Jones Industrial Average had increased in value tenfold. People were borrowing money to buy stocks. But on October 24, the market faltered and panic selling began. On Black Tuesday, about 16 million shares were traded, a record that would not be broken in forty years. The market had lost over $30 billion in two days, 25 percent of its value. The Great Depression that began with the stock market crash would last for ten years and would devastate the national economy. The Roaring Twenties were over.

Over the next five or six years, there was a chain of events that affected everyone in the country. First, the stock market called in the massive debts owed by speculative stock market investors who had invested more on credit than in cash, sometimes ten times more. There were thousands who lost everything, and although the traditional image of speculators jumping from skyscraper windows was somewhat exaggerated, there was a grain of truth in it. The Dow Jones Industrial Average, from 1929 to 1932, dropped from 260.64 to 41.22, its lowest level in the 20th century. Major manufacturers were hit hard by the stock market crash. General Motors' stock fell from $73 in 1929 to $8 in 1932. In the same period, RCA stock fell from $549 to practically zero. Most people, even those who were not invested in the market, were frightened by the state of the economy, and simply stopped or dramatically reduced spending on goods and services.

This, of course, reduced consumer demand, so manufacturers cut production and laid off workers. By 1932, manufacturing and sales were down 50 percent, and unemployment rose to 25 percent. American exports declined from $5.2 billion to $1.7 billion. Only as the Depression hit did many more manufacturers begin to realize that they had to do something, anything, to survive. Many were convinced of what they needed to do by Calkins and Holden advertising agency's 1932 book, *What Consumer Engineering Really Is*. Unfortunately, what Calkins erroneously called "consumer engineering" was actually more accurately called "industrial design," described earlier. Nevertheless, in his book, Calkins astutely and accurately summarized the current state of the economy:

> Many stopped buying while still able financially to continue, and many are still able but restrained by fear and misplaced thrift. The sudden cessation of buying slowed up the entire industrial machine. Retail storekeepers curtailed orders to factories. Factories cut down production, reduced wages, laid off men, still further reducing the number of customers for goods. In

> a comparatively short time, with all the resources of the country still intact, we had Depression. There were no longer enough buyers for the large quantities of goods we had learned to make and distribute so abundantly.[13]

Calkins recognized what many manufacturers did not: People had money to spend but they were just afraid to because of the rotten economy. Calkins also had the perfect solution to restore prosperity: attract buyers with new, innovative, and stylish products. He called it *obsoletism*: exactly what industrial designers had been doing for the last several years:

> Obsoletism is another device for stimulating consumption. The element of style is a consideration in buying many things. Clothes go out of style and are replaced long before they are worn out. That principle extends to other products—motorcars, bathrooms, radios, foods, refrigerators, furniture. People are persuaded to abandon the old and buy the new to be up-to-date, to have the right and correct thing. Does there seem to be a sad waste in the process? Not at all. Wearing things out does not increase prosperity, but buying things does. Thrift in the industrial society in which we now live consists of keeping all the factories busy. Any plan that increases consumption of goods is justifiable if we believe that prosperity is a desirable thing. If we do not, we can turn back the page to earlier and more primitive times when people got along with little and made everything last as long as possible. We have built up a complicated industrial machine and we must go on with it, or throw everything in reverse and go backward.[14]

As more and more manufacturers realized that to sell more products they had to invest in new designs to attract buyers, they sought more designers who specialized in that field, industrial designers, but those few practitioners were all overloaded already, and so more designers obligingly entered the attractive and lucrative new industrial design field.

It was the worst of times, as the Depression had reached its lowest point. Unemployment was at 25 percent, two million were homeless, 40 percent of all banks had gone bankrupt with panic runs, and many stores required cash purchases only. Those with money hung on to it for dear life. The presidential election in 1932 had resulted in the election of Franklin D. Roosevelt, who gave his inaugural address on March 4, 1933. He began with a phrase that echoed the sentiments of Elmo Calkins the year before: that fear of bad economic times had prevented the population from spending, which, in turn, crippled the economy. Roosevelt began: "So, first of all, let me assert my firm belief that the only thing we have to fear is ... fear itself—nameless, unreasoning, unjustified terror which paralyzes needed efforts to convert retreat into advance.... This is no unsolvable problem if we face it wisely and courageously."[15] Roosevelt provided the spark of hope and encouragement the nation needed at the time, often announcing new legislation to the public with informal "fireside chats" on the radio. Over the next few years, he would initiate the New Deal, an avalanche of federal legislation to combat the Depression. Legislation closed all banks briefly and reopened only those that were sound, providing federal loans if needed. Other legislation regulated the stock market to avoid future crashes, provided work for millions through federal infrastructure projects such as the Works Projects Administration (WPA), initiated rural electrification, and provided relief to farmers. New legislation gave more power to unions, repealed prohibition, and provided guaranteed retirement for the elderly through Social Security.

Meanwhile, despite the Depression, there continued to be advances in the refrigeration industry. Between 1928 and 1930, the House of Representatives, the Senate, the White House, the Executive Office Building and the Department of Commerce were all equipped with air conditioning, as were many important buildings across the United States. In 1929, Frigidaire introduced the first room cooler that used sulfur dioxide refrigerant and had a capacity of one ton. Because of its size, it was designed to be located outside the house or in the basement. In 1930, Electrolux introduced the first built-in refrigerator, and in 1931, the first refrigerator that was cooled by air. In 1930 Frigidaire introduced the first Hydrator, a humidity drawer for fruits and vegetables, with an adjustable slot to regulate the amount of air allowed in.

In the early 1930s, where could consumers buy refrigerators? Many stores that dealt with automotive products, such as the Western Auto chain and Firestone, diversified into home appliances and featured displays of major refrigerator manufacturers. Also, electric utility companies sold refrigerators and other electric appliances to encourage the use of electricity. Given the Depression, of course, sales were not very good.

In 1931, Frigidaire marketed its Hot-Kold year-round, central-air-conditioning system for homes. That same year, Southern California Edison Company installed an 800-horsepower, 480-ton vapor-compression air conditioning system in its Los Angeles office building; it provided summer cooling as well as heating in winter when the outside temperature was above 42°F. Both systems used what was now being called a heat pump. What, exactly, is a heat pump? Technically, the term "heat pump" is a generic term for any device that provides heat energy from a source of heat to a destination called a "heat sink." This is exactly what refrigerators, freezers, and air conditioners normally do, because they transfer

General Electric display in booth of the Pacific Tire Company at 1930 appliance show.

heat from warm spaces, making those spaces cooler, to other spaces (heat sinks), making them warmer. But when a heat pump is used for heating, it employs the same refrigeration-type cycle as in an air conditioner or refrigerator, but in an opposite direction—releasing heat into the home environment, drawing heat from the cooler outside air or from the ground. This, of course, can be much less costly than providing heat by gas or oil furnaces or electricity. So, at the time, to distinguish devices that only cooled (refrigerators, freezers, and air conditioners) from those that also heated, the term "heat pump" began to be used for those that provided heat.

This was not a new concept in the 1930s. Oliver Evans, who proposed the closed vapor-compression refrigerator cycle in 1805, had also noted: "Thus it appears possible to extract the latent heat from cold water and to apply it to boil other water."[16] In other words, his refrigeration concept could also be used to heat, as well as cool, simply by reversing the process. By 1852, William Thompson (Lord Kelvin) had proposed that an air-cycle refrigeration system be used to heat or cool the air in buildings, and described the design of such a machine. But the actual construction of such a machine did not occur until Austrian Peter Ritter von Rittinger (1811–1872) built such devices in Austria.

Von Rittinger published an article in 1855, "Theoretical and Practical Treatise on a New Method of Evaporating All Kinds of Liquids, Based on a Hydro-Mechanical Power System Applied to a Continuous Vapor Circulation with Special Application to the Salt-Brine Evaporating Process." Von Rittinger states in his introduction:

> Steam can generate mechanical work. Most of today's industrial advances are based on this fact. Physicists have no objections to also accepting that mechanical work can generate steam. As far as I know, nobody has tried to take advantage of this fact for the benefit of industry on a large scale. This treatise explores the possibility.[17]

Von Rittinger in 1856 constructed an experimental heat pump with an input of 11 kilowatts that was said to be 80 percent efficient. It was to be used to evaporate salt brine at Ebenee in upper Austria. But while it worked fine with fresh water, it failed to work with salt brine due to technical problems, having to do with the crust formed from the evaporation. Another salt plant based on Von Ritteninger's ideas was successfully built between 1870 and 1880 at Lausanne, Switzerland. In 1886, a discussion of heat pumps concluded that the most suitable such system would be one using ammonia vapor compression.

The most extensive work on heat pumps was conducted by Thomas Graeme Nelson Haldane, known better as Graeme Haldane (1897–1981) of Scotland beginning in the mid–1920s. He published his work in 1930 after testing a system installed in his home, suggesting that the vapor-compression system refrigeration cycle could provide a more economical method to heat buildings and swimming pools, as well as being used for central cooling and ice making.[18] This appeared to trigger a rash of heat pump development, because there was much effort to apply refrigeration systems for combined heating and cooling in the 1930s, including the 1931 Frigidaire Hot-Kold year-round central air-conditioning system for homes, and the Southern California Edison Company's air-conditioning system, both mentioned above. In 1932, General Electric engineers published an article discussing the theory, technical aspects, and economics of heat pumps as well as listing the reasons that they had not been commercially developed:

> No suitable refrigerating machine with the necessary characteristics of safety, quietness, lack of vibration, freedom from service, and high

> efficiency has been available commercially in sizes of 5 to 25 horsepower.
>
> The cost of electricity in most localities has been too high until recently to make this method of heating compare favorably with existing models.
>
> Very little has been known of the actual operating costs of a refrigerating system because of the special nature of refrigerators and auxiliary equipment needed, and the careful study of climatic conditions which is required to predict their performance.
>
> Very little has been known of the first cost of equipment for heating and cooling electrically because of the special nature of the equipment.
>
> People have not yet been educated to the needs of air conditioning, including refrigeration. Until they demand cooling equipment, there is little or no advantage of supplying refrigerating equipment for heating only.[19]

Some of these concerns would be met during the 1930s, and research and experiments continued, but not from General Electric. The first successful package heat pump was apparently the design of Henry L. Galson, Henry Heller, Charles Neeson, and Hans Steinfeld, produced by the De La Vergne Division of Baldwin-Southwark Corporation. U.S. Patent 2,130,327 was filed for it in 1932. It would be introduced to the market in 1933 and was of an "air-to-air" design rated at 14,000 BTU (British Thermal Units). It used an opposed-piston design, with a 1½-hp hermetic compressor with the newly available chlorofluorocarbon (CFC) refrigerant R-12. Due to the 750-pound weight of the unit, a duct connecting to the outside had to be provided for installation. About 100 units were produced.

Another installation by the Atlantic City Electric Company would be tested at low outdoor temperatures during 1934–1935. This system required four refrigeration units to meet the needed capacity. Despite these technical advances in heat pump technology, these systems were never very popular because they had to compete with fossil-fuel heating systems, which were becoming increasingly more reliable and less costly. By 1940, there would be only about 20 commercial installations of heat pumps.[20] The heat pump market would grow very slowly until the 1960s, when lower costs would make it affordable.

Meanwhile, in the mainstream air conditioning industry, equipment was shrinking in size, making it feasible for new applications. With the advent of Carrier's centrifugal chillers in 1924, smaller air conditioners became feasible for trains. In 1930 the Baltimore and Ohio Railroad tested a unit (U.S. Patent 1,982,125, *Air Distribution System and Apparatus for Railroad Cars*), designed by Alfred E. Stacey, Jr., and Herman Richard Arf, of Carrier Engineering Corporation, in the *Martha Washington* dining car on the *Columbian*, running between Washington and New York. To test the system, the car was first heated to 93°F, then the heat was turned off and the air conditioner turned on. Within 20 minutes, the temperature in the dining car was a comfortable 73°F. By 1932, the same railroad operated the first overnight train that was totally air-conditioned, the *George Washington*, between the two cities.[21] In 1931, H.H. Schultz and J.Q. Sherman invented an individual air conditioner that sat on a window ledge. The units were available for purchase a year later and were only enjoyed by people least likely to work up a sweat—the wealthy. The large cooling systems cost $10,000 to $50,000 ($120,000 to $600,000 in today's dollars).[22]

Also in 1932, the Gibson Company, which had been making wooden iceboxes since 1877, now began to manufacture electric refrigerators. They were following the example of other wooden cabinetmakers such as Leonard's, which provided cabinets for the early Kelvinators in 1914, or the Guardian Company in Detroit in 1916, which was bought by General Motors in 1918 and which became Frigidaire.[23] The Leonard brand was still being made by Frigidaire. The 1932 Leonard had a new foot-controlled door latch, called the Len-A-Dor, specifically designed to aid the busy housewife with both hands full of dishes. Competition in the new field was fierce; only those who made use of the economies of scale, mass production, and large advertising budgets succeeded. In 1932, the journal *Electric Refrigeration News* published a list of extinct household refrigeration makers. By then 111 firms had quit the business, 41 could not be located by mail, and 24 had been taken over by other manufacturers.[24]

In the late 1920s, isolated incidents of toxic and unstable refrigerants escaping were plaguing the refrigeration industry. Charles Kettering (1876–1958) at GM formed a research team to find a replacement for the dangerous refrigerants then in use. Thomas Midgely, Jr. (1889–1944), also of GM, headed the team that included Albert Henne and Robert McNary, and in 1928 the team improved the synthesis of CFCs and demonstrated their usefulness in refrigerants because of their stability and non-toxicity. Kettering patented a refrigerating apparatus to use the gas and issued it to Frigidaire, a wholly owned subsidiary of General Motors. In 1930, General Motors and DuPont formed Kinetic Chemicals to produce Freon (dichlorodifluoromethane) also known as "Freon-12," "R-12," or "CFC-12." Frigidaire was the first manufacturer to use Freon. Soon Freon would replace older refrigerants used by many refrigerator manufacturers, but others continued to use the older refrigerants in new systems into the 1940s. The use of CFCs was interrupted by World War II, when its production was co-opted for the war effort.[25] Eventually, CFCs would be banned in the 1980s due to their harm to the world's ozone layer, but no one in the early 1930s knew of this danger.

General Electric in 1930 published the first known complete major appliance service manual for users, which gave full instructions on servicing its Monitor Top refrigerator. GE also made the first important breakthrough in insulating refrigerators in 1931 with the development and production of Thermocraft insulation, which reduced the weight of insulating materials used in refrigerators from 12 pounds per cubic foot to 2.5 pounds. It also reduced leakage in the cabinets by 15 percent. For years, competitors paid GE royalties to use Thermocraft.[26]

At about this time, refrigerator manufacturers, with their sales suffering severely in the Depression, discovered industrial designers, who were already designing automobiles and other "artless" industry products, such as washing machines, home appliances, and business equipment. There were only a dozen or so of them at the time, and most were extremely busy. In 1930, Westinghouse hired its first industrial designer, a design instructor from the Carnegie Institute of Technology in Pittsburgh, Donald Dohner (1892–1943), as "director of art"; he would design all its products from locomotives to ashtrays until 1934. Among these were refrigerators, coolers, and refrigerated water fountains. Before Dohner, Westinghouse refrigerators were of fairly traditional designs, not very different from those of its competitors, but it is interesting to see some of the prototypes that Dohner designed but were never produced. One in particular was a concept that tried to address the typical problem of the difficulty of user access to food far in the back of a deep food cabinet. It is a typical example of the type of innovative thinking brought to industry by industrial designers.

The concept was called the "rotary cabinet design." The cabinet was only about half of the depth of ordinary refrigerators. Rather than opening with hinges like other refrigerators, the door was actually a fixed part of a rotating food-cooling compartment. When one side of the door was pushed, and rotated 180 degrees, lazy–Susan fashion, it brought all the food within easy reach, as the rounded shelving in the rear protruded several inches beyond the main cabinet in the front, but the depth was only about a foot from the front of the shelf to the rear. Further, the small freezer compartment was a separate, smaller door above the rotating door, which, when the knob was pulled, pivoted downward and outward to access the ice cubes. The drawback to the design, of course, was that while its height and width were the same as typical refrigerators, its depth was less than half, so its capacity was far smaller than competitors, the prob-

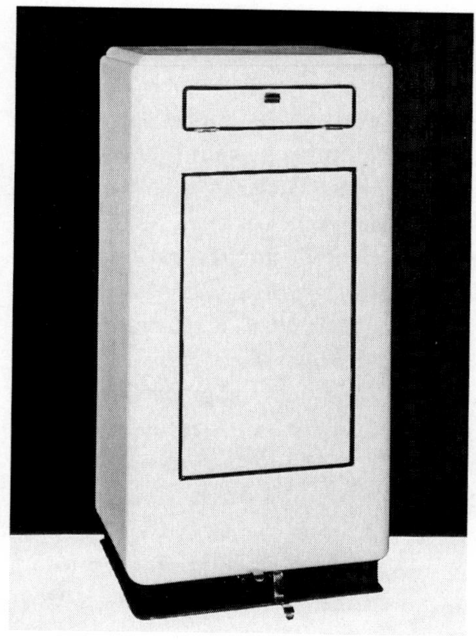

Above, left: ***Westinghouse rotary cabinet concept, closed.*** **Above, right:** ***Westinghouse rotary cabinet concept, partly open.*** **Right:** ***Westinghouse rotary cabinet concept, open, all by Donald Dohner (all courtesy Hampton Wayt).***

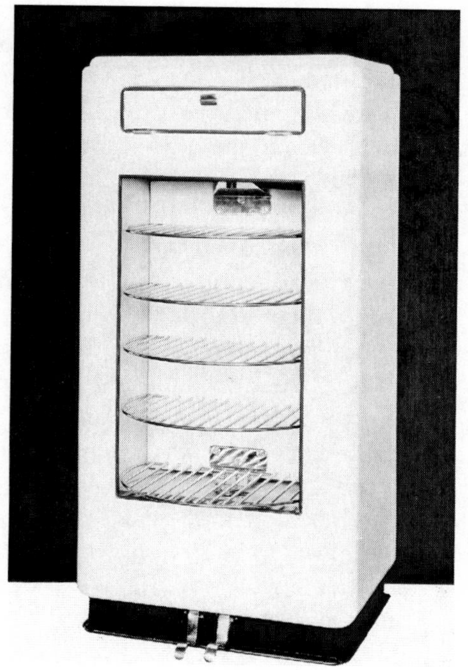

able reason it was never produced. Capacity was a big competitive sales feature at that time. The concept of rotating shelves for access in the rear of the food cabinet would not actually appear on the market until the mid–1950s.

Westinghouse began making electric coolers for Coca-Cola in 1930, but also continued to make traditional ice coolers. Coca-Cola had been using ice coolers since it began selling soda pop at the end of the 19th century. They were initially round coolers made from wooden half-barrels with wooden lids and the words "Drink Coca-Cola in Bottles" stenciled on the side. Customers simply lifted the tub's lid, took a bottle, and then paid the clerk for their drink. In 1910, a Georgia bottler named George Cobb had built the first coin-operated vending cooler for Coca-Cola, but it only held 12 bottles. By the 1920s, numerous companies with names such as Freez a Bottle, Icebergdip, and Walrus were producing Coca-Cola–branded coolers for retailers. Other round, ice-cooled coolers made by Icy-O were made of metal. Icy-O made both floor and countertop models that held from 72 to 120 bottles; some featured a compartment for empties, which were cleaned and refilled. In 1928, Coca-Cola hired a sheet-metal manufacturing firm called Glascock Bros. to design and build a cooler that it could sell to retailers. By 1929, the rectangular Glascock cooler, with an icebox above and angled racks for empties below, was ready. Coca-Cola sold 32,000 of them at $12.50 each. In comparison, Westinghouse's 1930 electric model cost $150, but its traditional coin-operated Vendo Top cooler still gave customers an ice-cold Coke for a nickel.

Of course, by now, those lucky people who had electric refrigerators did not have to go to the local store or gas station for a Coke; they could enjoy cold Cokes at home any time they wanted. They just had to pop a few ice cubes into a glass of Coke and enjoy, right? Well, not always. One of the most frustrating problems of early refrigerators was the ice cube tray, which froze neat little cubes of ice separated in the tray by thin aluminum fins. If one wanted a few cubes for a drink, one had to remove the entire tray and take it to the sink to douse it with hot water to loosen the cubes. Often, they still stubbornly clung to the fins and the tray itself, and one had to resort to banging the tray, upside down, against the edge of the sink to dislodge the individual cubes into the sink and pick up the one or two needed.

Then one was left with a dozen unused cubes. What to do with them? Throw them in the sink to melt away, and refill the tray with water to re-freeze? Put them back in the tray, carefully arranging them individually to fit in their original positions within the aluminum fins? Was all this effort worth just several cubes? It was, but there had to be a better way, and in fact, there was. In the late 1920s, engineer Lloyd Copeman introduced a rubber tray, which could be twisted to remove the ice. In 1933, Guy L. Tinkham, then vice president of the General Utilities Manufacturing Company in Detroit, which produced household appliances, invented the first flexible, stainless steel, all-metal ice cube tray with square depressions for ice, notched at each side, which flexed sidewise to eject the cubes (U.S. Patent 1,997,839, *Method of Making Ice Trays*, granted 1935). The

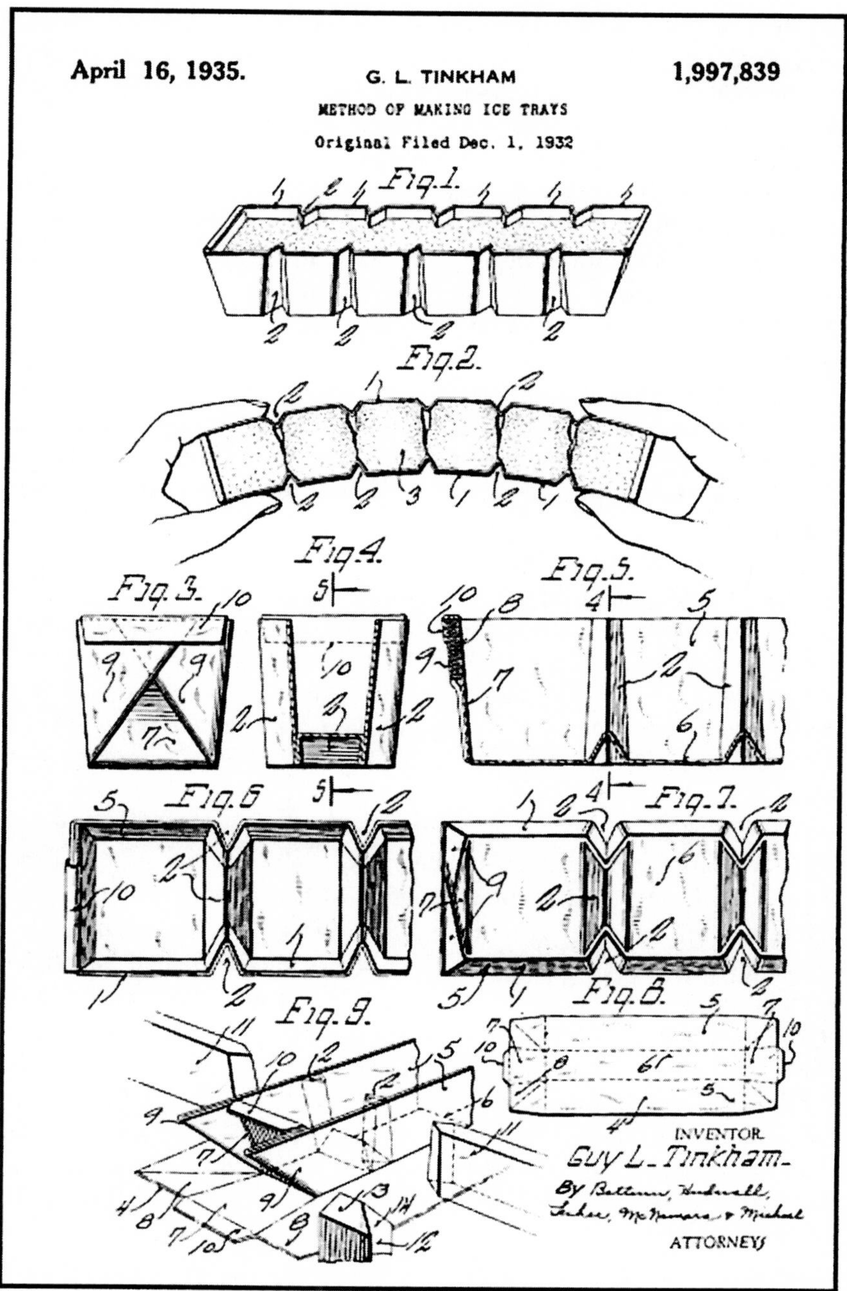

Tinkham ice tray, U.S. Patent No. 1,997,839, 1935.

tray held only six ice-cube depressions in a single row. Flexing the tray sideways cracked the ice into cubes corresponding to the division points, or side notches, in the tray, and then forced the cubes up and out. Pressure forcing the ice out was due to the 5-degree draft (slope) on all sides of the cube depressions. Tinkham also invented and patented the equipment used to fabricate the trays. Tinkham's invention was named the McCord Ice Tray and cost $0.50 in 1933. Later, various designs based on the McCord concept would be developed, including flexible plastic trays that functioned in a similar manner to the 1920 rubber trays.

There were, however, more serious problems in the world than ice cubes sticking in a tray. In 1933, Adolph Hitler came to power in Germany, initiating the potential for another world war. As mentioned earlier, scientists such as Albert Einstein began fleeing from Nazi persecution and military buildup, and brought with them technological knowledge helpful to America. Most Americans were too preoccupied with economic fears and problems to pay much attention to what was happening in Europe. They didn't need disturbing news from Europe—they needed some good news at home. The World's Fair in Chicago filled that need.

The Century of Progress Exposition in Chicago, which opened in the summer of 1933 and attracted 48 million visitors, lifted public spirits. It was a showcase of new technology, transportation, architecture and modern design. Several streamlined, lightweight trains made their debut at the exposition, along with streamlined experimental automobiles, both inspired by the speed-inspired forms evolving in the booming aircraft industry. By 1932, GM had introduced streamlined cars to the market, when automotive production had declined to about a quarter of its pre–Depression levels. As described earlier, GM had established an internal industrial design (or styling) group in 1927, and had led the way in modern design for the "artless" industries. In 1930, Westinghouse had established a similar internal corporate design department, and in 1933, marketed the first completely self-contained room air conditioner using Freon–12. In 1932 Chrysler established an internal design department, and in 1933, the Radio Corporation of America (RCA) initiated one as well. Major companies were embracing industrial design, particularly that who were manufacturing refrigerators. Designers would modernize them in a manner that set the appearance standards that still exist today: simplicity, modernity, and user friendliness.

Chapter 7

The Design Decade

Designers were heavily involved with most of the exhibits at the Chicago Exposition, and after the fair, many entered the rapidly expanding industrial design field. One of these was Raymond E. Patten (1897–1948), who had designed exhibits for the Edison General Electric Company in Chicago. In 1933, GE established an internal design group at its corporate headquarters in Bridgeport, Connecticut, and named Patten as its head. Under his leadership, GE engaged one of the most prominent industrial designers, Henry Dreyfuss, to redesign the 1927 Monitor Top refrigerator, which, as you may recall, had the refrigeration equipment exposed on the top of the food compartment in a large, mechanical-looking cylinder. Dreyfuss had earlier in 1933 designed a Toperator washing machine for Sears, Roebuck and Company. The Toperator name referred to the fact that Dreyfuss had relocated all the machine's controls on top, making them more accessible to users.

Dreyfuss completely redesigned the Monitor Top by relocating the refrigeration and motor equipment under the food compartment, totally concealed from view. Not incidentally, this raised the food compartment higher and made it more easily accessible to users. The new design was called the Flat-Top, to distinguish it from the old Monitor Top and it debuted in 1934. A decorative, embossed panel about six inches wide ran from top to bottom of the front. Such decorative linear panels were common features of modern Art Deco design. The Flat-Top was produced under both the GE and Hotpoint brand names. That same year, Hotpoint was integrated within the GE organization. Dreyfuss also designed a modified version of the original Monitor Top and at about this time, GE introduced the first adjustable shelves in refrigerators.

The Flat-Top design was a microcosm of the transition between traditional design and modern design by industrial designers, the process taking place during the 1930s that significantly helped manufacturers in the "artless" industries reach their objective of low consumer cost through high volume mass production, and allowing millions in the middle class to enjoy the fruits of technology. The manufacturing method for refrigerators was constantly changing during the 1930s, in order to produce higher quantities in shorter time and at lower cost. A prominent industrial designer, Harold Van Doren (1895–1957), in 1949 described these progressive manufacturing procedures:

Hotpoint (General Electric) Flat-Top refrigerator by Henry Dreyfuss, 1933 (courtesy the Museum of Innovation and Science, formerly the Schenectady Museum, Schenectady, New York).

[Initially,] cabinet exteriors consisted of flat sheets of vitreous-enameled steel, fastened with screws to a wooden frame. Up until 1933, [the cabinet design of] electric refrigerators mimicked the wooden icebox except that it was made of sheet steel and painted white. The main body of the cabinet was a wrap-around from front to back [bending sharp 90-degree corners on a brake press]. A large opening was pierced in the front to receive the vitreous-enameled food compartment. The whole was surmounted with a shallow drawn lid, often formed to imitate wood molding.

[In 1933] the Norge Company engaged an industrial designer, [Lurelle] V.A. Guild [1898–1985]. Following his appointment, Norge produced several models of sizes in a new form. The cabinet was produced in sections from large metal dies. Each side was made in one piece, which, together with a pressed top member closely matching the conformation of the side panels, was fastened to a skele-

> tonized frame. The entire machine was then put together on a progressive assembly line. [This was the same process used for the GE Flat Top.]
> From 1936 on, as production began to climb over the two-million-unit mark, few manufacturers could remain in competition if they retained the old methods. Furthermore, the public liked the smooth, easily cleaned, nearly joint-less surfaces.... Westinghouse adapted the "tangent bender" ... to the manufacture of refrigerator cabinets. In this method the entire shell of the refrigerator is formed from one long piece of steel wrapped up one side, across the top, and down the other side. The finished piece forms an inverted "U." The radii at the upper corners *cannot be less than 3½ inches.* The saving over the previous method of assembly by panels is ... that no separate inner frame is required. Below center a transverse pan is welded into the "U." Another pan ... is welded in at the bottom [on which] the compressor and condenser are mounted. Meanwhile, the food storage compartment ... has been separately fabricated ... and passed through the vitreous enameling process. This ... is inserted into the shell, and blocks of insulating material are slid in at the sides, top, and bottom of the cavity. When the evaporator has been installed and connected with the compressor-condenser unit, back insulation is inserted and a back panel secured in place.[1]

The last process that Van Doren described inevitably caused the corners on refrigerators to be rounded, rather than sharp, a manufacturing reality that had to be accepted by industrial designers. But purist art critics blamed them for streamlining objects that were not in motion, an accusation that stuck for years and put industrial designers on the defensive.

In traditional "artless" industries, including radios, bathtubs, stoves, sinks, refrigerators, all had legs, because engineers generally regarded them as furniture, which traditionally had such legs. They were sometimes also cosmetically decorated with token artistic patterns, such as traditional leaves or flowers. Nevertheless, engineers were proud of their creations and saw no need to conceal their functional, ingenious mechanisms from consumers, no matter how disorganized their appearance, or how detrimental to user convenience they might be. Thus, the Monitor Top refrigerator featured its refrigeration equipment at eye level of the user, forcing the user to stoop in order to access the food storage compartment. To engineers, the primary mechanical function of the product (the compressor) was its most important visual feature.

Industrial designers were in direct conflict with this kind of thinking. They saw products not as traditional furniture but as new household appliances freed of the past and designed for the future. Legs were not needed for refrigerators, and industrial designers soon eliminated them not only from refrigerators, but also from stoves, beds, and other furniture, including console radios. To industrial designers, the most important feature of the product was that which inspired sales—the external appearance and how it looked in the home. Industrial designers' logic was that in a competitive sales environment where all products were essentially equal in function, the buyer would choose the one that best fit his or her home décor, or the one that looked the best or most modern. Designers sought to cover the mechanical components that engineers loved and to simplify the exterior with the simple, clean, geometric lines of modern design. Dreyfuss concealed the refrigerating equipment on the Flat Top below the food storage compartment, which raised the access to food closer to eye and hand-level convenience.

Understandably, this conflict of perceptions between engineers and industrial designers resulted in disagreements as the new designs evolved. Each group regarded the other as a threat to its own strong philosophy. In addition, engineers regarded designers as artists who knew nothing about the electrical and mechanical worlds, and designers saw engineers as ignorant and dismissive in the visual arts and consumer psychology. Finally, engineers resented the exorbitant fees demanded by industrial designers.

Engineers soon learned their lesson in the most spectacular design failure in history. Although Chrysler had established its internal design department in 1932, it was dominated by engineers, and unlike GM, where stylists were given full reign to shape the exterior form of cars, Chrysler permitted its stylists only to design cosmetic features such as decorative Art Deco trim for its highly touted, streamlined, 1934 Chrysler Airflow. Chrysler's best engineers, who shared the premise with many automotive engineers that functional streamlining alone would make cars more beautiful, designed the external form as well as the mechanical components.

The Airflow was indeed a marvel of innovative engineering features such as lightweight but strong unitary construction, more interior space, and a smoother ride. Chrysler even called in famous industrial designer Norman Bel Geddes as a token industrial designer to promote the car and lend his name to its introduction. But the styling was quite different from other cars—too different. To many, it was bulbous, awkward looking, and ugly. Car buyers avoided it like the plague, and only sales of Chrysler's more conventional cars saved the company. In desperation, Chrysler put well-known automotive designer Raymond Dietrich (1894–1980) in charge of styling, and he added a new hood to make it more conventional, but these minor cosmetics failed to turn things around. Engineers everywhere had to admit that industrial designers were absolutely required to design products that would appeal to the public. The Airflow was the last car to be styled by engineers.

Meanwhile, in the refrigerator industry, Powel Crosley, the refrigerator manufacturer, had in 1932 patented the idea of shelves in refrigerator doors, an innovative feature of outstanding convenience to users. In 1934 his Shelvador refrigerators were introduced; over the next few years, they became one of his best selling models.[2] Designed by famous industrial designer Walter Dorwin Teague,[3] the innovative feature left competitors scrambling. After Crosley patents expired in 1949, all refrigerator manufacturers would universally adopt this door-shelving feature.[4] Other refrigerator manufacturers soon followed GE's redesign of its iconic Monitor Top by Dreyfuss, the Flat-Top. In 1934, Norman Bel Geddes designed refrigerators for Electrolux/Servel in modern styles. Also in 1934, Kelvinator introduced the first two-door household refrigerator. Major manufacturers were already aware of industrial design, but soon, all manufacturers would get the message.

In the February 1934 issue of *Fortune* magazine, there appeared a highly influential article that finally described the accomplishments of industrial designers for the benefit of the entire business community. It was titled "Both Fish and Fowl," referring to the little-known fact (by many manufacturers) that industrial designers were both artists *and* businessmen. The article was written anonymously but was later known to be by George Nelson (1908–1986), winner of the 1932 Rome Prize in Architecture and assistant editor of *Architectural Forum* magazine at the time. Nelson would after World War II become head of industrial design for the Herman Miller Furniture Company. Nelson began his *Fortune* article with a quote from Oliver M.W. Sprague, a retired economic advisor to President Roosevelt in 1933. Sprague placed the blame for the continuing Depression squarely on manufacturers: "Failure of industries to adapt policies designed to open up additional demands for industrial products, is, in my judgment, the chief cause of the persistence of the Depression."[5] This confirmed the early warning by advertiser Elmo Calkins two years before, who advocated the improvement in styling by manufacturers to increase demand for products with what he called "obsoletism," a process of continuous redesign to meet customer desire for the "latest styles." And although a few manufacturers had heeded his warning, many still were not aware of how to do this, or were blissfully unaware of the problem. Nelson's article told readers exactly what industrial design was, and gave specific examples of how it improved sales. The article popularized the term "industrial design," clearly separating these designers from the many

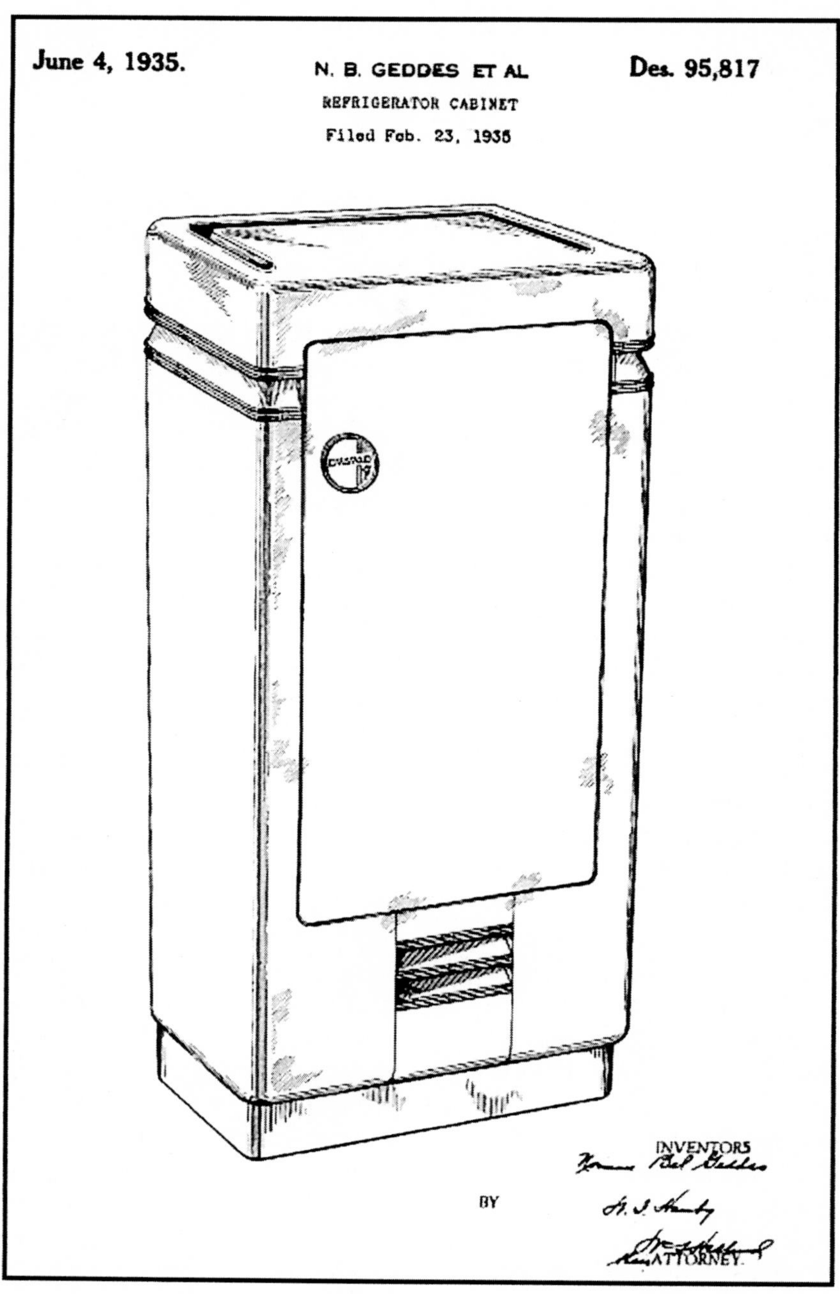

Servel refrigerator by Norman Bel Geddes, U.S. Patent No. D95,817, 1934.

who were in the "art industries" of decorative products. Industrial designers analyzed the product and the market, and made extensive changes in form and configurations, not just cosmetic additions. Although there were at the time only several dozen industrial designers, Nelson described a dozen or so of the most successful in great detail: what they designed, how they worked, and what they were paid. Photos of them and their successful designs were included.

Some were already well-known, such as Henry Dreyfuss, Raymond Loewy, and Walter Dorwin Teague. Within a few years, they would become public celebrities, and like the automotive companies of General Motors, Chrysler, and Ford, would become the "Big Three" of industrial design. Others, such as Lurelle Guild, Harold Van Doren, John Vassos, Joseph Platt, Gustav Jensen, Russel Wright, Donald Dohner, Jo Sinel, and George Sakier, were relatively unknown at the time but would also become quite successful. The most dramatic part of the Nelson article was the citing of sales increases as a result of specific industrial designs, which ranged from 25 percent to 900 percent, incredible figures for Depression times. Their fees were also quite incredible. Annual retainer fees were as high as $50,000 and hourly rates of $50, when corporate executives made about $5,000 annually, and common laborers 50 cents per hour.

But in these grim times, manufacturers would pay anything for sales increases like these, and the floodgates opened. Demand for industrial designers skyrocketed, and many new professionals entered the field. Soon there would be hundreds. Many who were already practicing increased their staff substantially to accept more clients. From here on, industrial design would become legitimized in industry and an essential component to success in the marketplace. That same year, 1934, Carnegie Tech (now Carnegie Mellon University) in Pittsburgh initiated the first degreed educational program in industrial design in the world, headed by Donald Dohner.

In 1934, Sears, Roebuck and Company initiated an in-house industrial design department, headed by John Morgan (1903–1986), a talented industrial designer. At the time, total refrigerator annual sales in the United States were 1.3 million units, compared to 200,000 in 1926, and retail prices had dropped to about $170. However, even before this, in 1932, Sears had gone to Henry Dreyfuss to request his redesign of its Coldspot refrigerator, originally designed in 1928 by one of its suppliers and introduced in 1931. At first the Coldspot had been a disaster. Production could not meet demand, and quality control was non-existent.[6] Sears knew there was a huge market in electric refrigerators but that costs

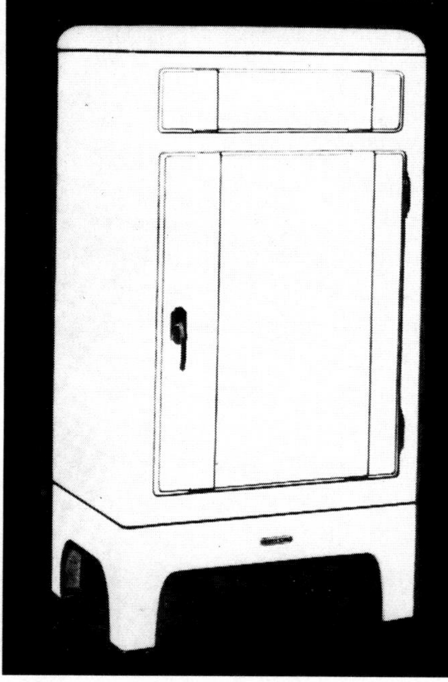

Above: ***1931 Coldspot refrigerator.*** **Below:** ***Raymond Loewy, ca. 1934 (courtesy Hampton Wayt; photograph, Hal Phyfe).***

prevented most people from buying them in the Depression. So Sears decided to design its own refrigerator. At the time, most refrigerator manufacturers were building 4-cubic-foot models. Sears wanted to build a 6-cubic-foot model and sell it at the same cost as a 4-cubic-foot model.

Henry Dreyfuss had to decline Sears's request, because he was currently working on the GE refrigerators. Industrial designers were ethically bound to not work on competitive products. So Dreyfuss recommended that Sears contact Raymond Loewy, which it did. When he accepted the job, Loewy described the original Coldspot as "ugly," an "ill-proportioned vertical shoebox 'decorated' with a maze of moldings, panels, etc., perched upon spindly legs high off the ground, and the latch was a pitiful piece of cheap hardware." Mechanically, it lacked the sealed compressor that was the standard since the 1927 GE Monitor Top.[7]

On Loewy's staff at the time was industrial designer Clare Hodgman (1911–1992) who did most of the physical development work at Sears (and after he finished the project would be hired by Sears). The new Coldspot Super Six design was developed meticulously with clay models, similar to automotive procedures, but slightly smaller than actual size, presumably under Loewy's supervision, and was introduced in 1935. It was advertised as "Full 6-cubic foot size" for "about half the usual price," and buyers were asked to "compare it with any other refrigerator of similar size, selling in the $250 to $350 class." The design suggested quality and simplicity. It was a sleek, white, vertical, rounded-top box with three vertical, parallel ribs running from top to bottom in the center of its front façade, accenting the rectangular chrome Coldspot name escutcheon at eye level, which was framed by the decorative ribs. The parallel rib theme was repeated on the lower pullout drawer for vegetables. The token legs were formed integrally with the base assembly, which lifted the unit about four inches above the floor.

Unlike the GE refrigerator by Dreyfuss, the Coldspot motor and condenser were on the top of the food compartment, but totally concealed under a removable panel, which blended in perfectly with the external shell. Loewy also developed the Touch-O-Matic latch, designed so that the housewife with both hands full could still open the refrigerator by pressing slightly with her elbow on a long, vertical bar that also served as a latch handle. "The latch was substantial as well as attractive (like the door handle of an expensive automobile)," according to Loewy.[8]

Shelving had been a serious problem in the refrigerator industry. It had to be formed of metal wire and assembled by hand, welded, and dipped in rust-proof finish. It rusted anyway, looked messy, and the cost was excessive because of the amount of labor involved. Loewy had been inspired in his design of the interior shelving pattern of the refrigerator by the front end radiator grille of the Hupmobile automobile on which he was working at the same time. It was fabricated with strips of perforated aluminum, reinforced on the back by longitudinal ribs. Loewy had samples cut to the dimensions of Coldspot shelves and showed them to Sears. They were impressed with a design that was inexpensive and would never rust.

Sears paid Loewy $2,500 for the design, but he had spent nearly three times as much in expenses. It would turn out to be a wise investment for him, nevertheless. Sales on the Coldspot Super Six doubled from the previous year to 59,000 after the new model was introduced in 1935.[9] A highly appreciative Sears paid Loewy $25,000 for three future model design changes for the Coldspot, each to be more modern looking than its predecessor.[10] In these three models, appearance changes would range from minor to extensive, but they would also include a number of user conveniences.

Electric refrigerator manufacturers advertised their latest models incessantly and dramatically. In 1935, a refrigerator was the star of the first commercial Technicolor film. The film ran an hour and a half and had Hollywood stars and a romantic comedy script rooted in a family's

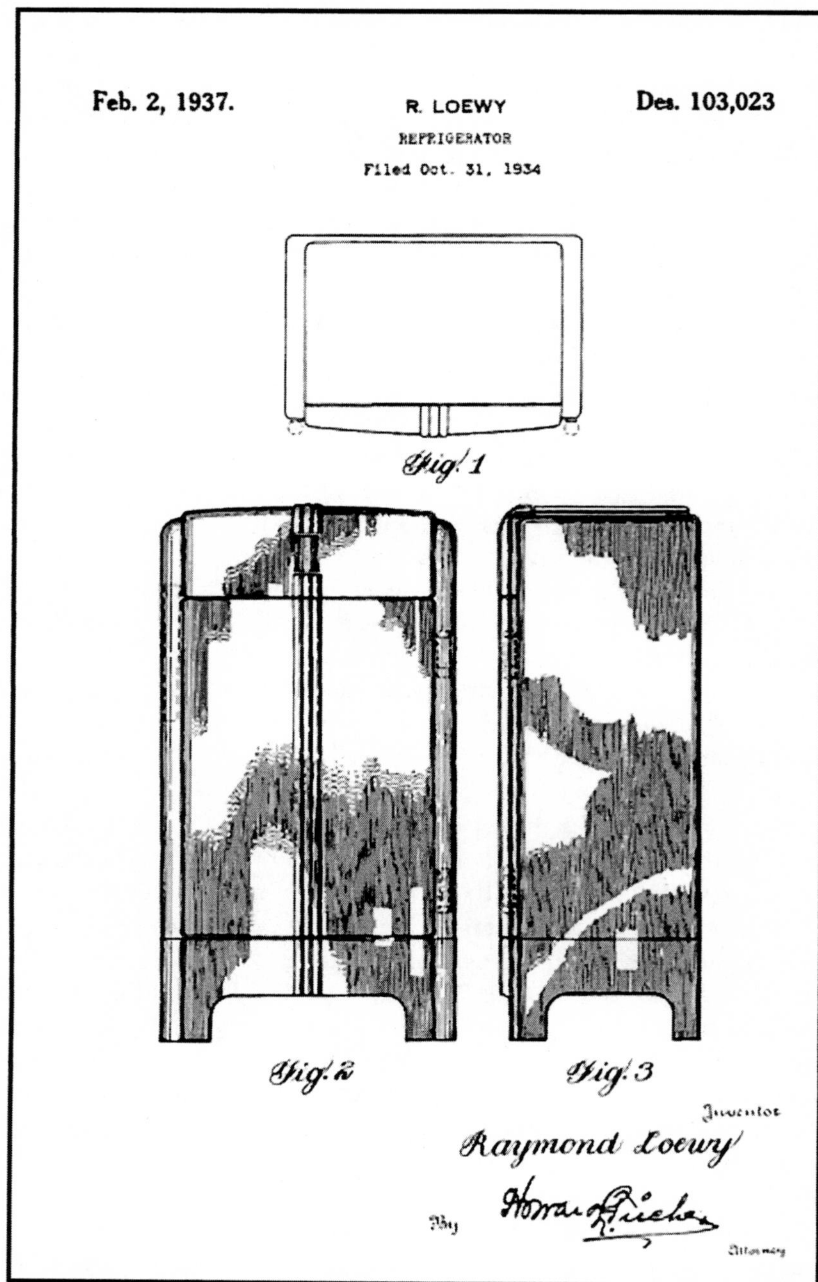
Feb. 2, 1937.
R. LOEWY
REFRIGERATOR
Filed Oct. 31, 1934
Des. 103,023
Fig. 1
Fig. 2
Fig. 3
Inventor
Raymond Loewy
By
Attorney

1935 Coldspot by Raymond Loewy.

need for a complete electric kitchen.[11] During the same year, commercial air transportation became realistic and profitable. The Douglas DC-3 airliner carried 21 passengers, and when United Airlines purchased the planes from Douglas for commercial flights, they engaged Henry Dreyfuss to design the interiors, which included the first air conditioning system for any aircraft. United claimed that their planes flew "three miles a minute" (180 miles per hour).

Air travel made it easier for Raymond Loewy to travel to Chicago, where he was already working on his next design for Sears' 1936 Coldspot, which was distinguished by a dominant vertical dark stripe down the center of the front façade. It could have been an added chrome or black strip, or possibly created by making each of the door and motor compartment covers in two separate metal stampings, but however it was made, it apparently was intended to replace the original vertical three ribs in the 1935 model as a central decorative feature. The lower Handi-Bin for vegetables had a half-spherical shaped pull handle, and included decorative legs that moved with the bin, beside the permanent functional legs that were integrated into the main body stampings. The door featured a large, circular Coldspot logo beside the Touch-O-Matic door latch, replacing the rectangular logo on the 1935 model.[12]

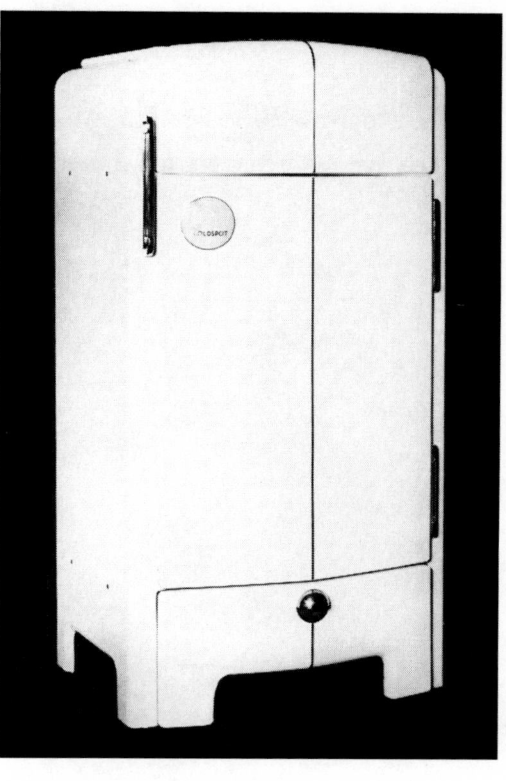

1936 Coldspot by Raymond Loewy.

The next Loewy design for Sears in 1937 was distinguished by a subtle vertical crease in the front façade from top to bottom, creating two planes instead of a flat surface, a much more elegant and subtle visual feature than the central vertical dark stripe between halves on the 1936 model. The hinges were virtually concealed, and the Coldspot escutcheon was a horizontal, rectangular plate above the door on the right side. There

Top: ***1937 Coldspot by Raymond Loewy.*** **Bottom:** ***1938 Coldspot by Raymond Loewy.***

was a Touch-O-Matic horizontal door handle, functioning similarly to the 1936 model. The lower Handi-Bin with hemispherical pull handle went all the way to the floor, with only the faint suggestion of legs in each lower corner.

Loewy's final Coldspot design of 1938 retained the vertical center crease of the 1937 model. He converted the round Coldspot escutcheon into a sculptured hemispherical recess, and the spherical end of a horizontal latch handle was centered in the recess, suggesting a bull's eye. The name Coldspot appeared in small letters above the Touch-O-Matic door handle. The one-foot-high empty space between the floor and the food compartment was a functional Handi-Bin with a spherical pull knob, which, like the end of the latch handle, was in the center of a semi-spherical recess.[13] The design still had a token suggestion of legs, which were camouflaged by narrow horizontal black strips alternating with white, suggesting the refrigerator was floating three inches off the floor. After that, refrigerator legs essentially disappeared completely from future designs in the industry. The Coldspot had a capacity of 8 cubic feet, weighed

473 pounds, and was priced at $149.50 ($2,470 in today's dollars). Coldspot annual sales of these later Loewy designs in 1936 would be 216,000, and by 1939, 290,000. Sears had climbed from 11th to 2nd place in refrigerator sales.

With the Coldspot, Loewy had made his personal reputation, and celebrated by relocating to a penthouse studio at Fifth Avenue and 47th Street in New York, and by opening his first office in London, England, the first U.S. industrial designer to go international. Loewy now had twelve clients and was doing very well.[14] His Coldspot refrigerator designs reverberated throughout the industry, and he was considered the leading refrigerator designer. In addition to providing an attractive modern exterior, Loewy was also developing manufacturing cost reductions and providing convenience features for users. Combined, these were the valuable design contributions that all industrial designers were providing to manufacturing clients. All increased sales, the fundamental purpose of the new profession. But industrial designer' most significant contribution was to provide women with appliances that were designed with them specifically in mind. Before, manufacturers and engineers designed products with only the mechanical performance in mind. Products were sold in the same way. Shelley Nickles in her article, "Preserving Women: Refrigerator Design as Social Progress in the 1930s," quotes Sam Vining, Westinghouse's director of department store sales, who in 1940 described a typical salesman of the early 1930s:

> He caught hold of the door handle, and he said, "Look, this door handle moves three ways." And the woman was supposed to drop dead. Now, what the hell difference does it make to the refrigerator how many ways that door handle moved? ... Fifty per cent of our business is preserving women, not fruit, and it makes a lot of difference to the woman whether she can walk over with her arms full of something and it will open in any direction she wants." But that salesman didn't tell her that—he was selling merchandise, not people.[15]

Nickles concludes:

> Consumers experienced new domestic technologies, and therefore the process of modernization, in large part through the mediating influence of industrial design. Design elements were related only tangentially to the technical function of the refrigerator as a food preservation device, but were central to its perceived social function of "preserving women" by aiding the servant-less housewife. Historians of household technology have asked how the refrigerator "got its hum" but not how it got its streamlined curve, vegetable drawer, or door handle.... The exploration of industrial design as part of a larger process of social interaction offers insight into

how new technologies were domesticated, on whose terms, and with what social consequences.[16]

But industrial designers were also the architects of the modern kitchen. They were contributing in a way that manufacturers of major appliances, or even consumers, had long failed to even recognize: their individual, freestanding appliances were just components of a larger environment, the kitchen. Industrial designers understood that principle, and designed appliances in cubical or rectangular forms (rectangular parallelepipeds, in geometrical terminology) not simply because it was the modern style, but also because these forms could be visually integrated into a kitchen that was becoming a total, functional interior design; a machine for food storage, preparation, and cleanup, as Christine Frederick had advocated in 1920.

Frank and Lillian Gilbreth, followers of Taylorism, had advanced these ideas of efficiency, and after Frank's death, Lillian partnered with the Brooklyn Borough Gas Company in 1929 to develop the Gilbreth's Kitchen

Gilbreth's Kitchen Practical, 1929.

Practical to showcase new gas-fueled appliances as well as Gilbreth's research on motion-saving in kitchens. Many kitchens in the 1930s were still far from being any kind of a functional "machine." They still had furniture and appliances of different shapes and designs, located willy-nilly and creating a disorganized and eclectic look, often without counters or cabinets. The new architectural forms of major appliances, primarily white stoves, refrigerators, and sinks, along with wall-to-wall and floor-to-ceiling kitchen cabinets for dishes, pots and pans, and spacious countertops, by industrial designers, were slowly creating a built-in look to the entire kitchen that was attractive, efficient, and functional as well, to the delight and appreciation of millions of housewives.

The trend toward more up-to-date kitchens got a big boost in 1935, when the National Kitchen Modernization Bureau was established jointly by the Edison Electric Institute (later the Electric Energy Association) and the National Electrical Manufacturers Association to actively promote kitchen modernization throughout the country. Tied in with the Federal Housing Bureau on general modernization of the home, the new bureau launched an extensive program that included the creation of model kitchen displays; radio programs; distribution of modern electric kitchen plan books; and a feature motion picture entitled *The Courage of Kay*, in which the subject was dramatized with numerous tie-ins with retailers, appliance and kitchen equipment manufacturers, builders, and others.[17]

This trend had been largely influenced by the design of the prefabricated and modular kitchens introduced in 1937 in the Netherlands by the Bruynzeel Company and designed by Dutch industrial designer Piet Zwart (1885–1977). He was engaged to design a modern kitchen for mass production, the result of which was highly progressive for the time. All-metal cabinets and fitted appliances were color coordinated and gave the kitchen an appearance of absolute visual unity and efficiency that is still typical of our kitchens today. Bruynzeel kitchens are still available, but now they have many competitors. Industrial designers in 1937 were well aware of these Bruynzeel kitchens and knew that they had to design individual appliances that could be inserted into such kitchens both spatially and visually. The era of the freestanding appliance, independent of all around it, was at an end, and industrial designers knew it. By the end of the 1930s, many of America's kitchens were well integrated, such as the Pardoe "continuous kitchen" by the Kitchen Equipment Company, which was inspired by the Gilbreth's Kitchen Practical design of 1929.

By this time, two million Americans owned mechanical refrigerators. In 1936, Kelvinator had added room air conditioners to its line and had merged in 1937 with Nash Motors, the automobile manufacturer, creating the Nash-Kelvinator Corporation. In 1938, Frigidaire debuted its Silent Meter-Miser refrigerator with the first Quickube, an aluminum ice cube tray with a built-in cube release, solving the maddening problem that had plagued users for decades: getting stubborn ice cubes out of frozen trays. The cube release was a 12-inch-long lever lying lengthwise along the center of the tray, which, when pulled up, pivoted at one end, and, being attached to all the individual movable fins separating the cubes, twisted them and loosened the cubes for easy removal. This was a variation of the 1933 McCord concept of a flexible tray, described earlier. The Quickube concept soon became standard in all refrigerators. That year, General Electric introduced a butter keeper. Heat controls allowed users to set their desired level of butter softness, but this electrical feature was not grounded and was known to shock people. Also in 1938, Frigidaire introduced its first air-cooled, window-type, air conditioner.

The Depression worsened in 1938 as the economy took a sharp downturn. Industrial production, including durable goods, fell almost 30 percent. Unemployment jumped from 14 percent in 1937 to 19 percent in 1938, rising from five million to twelve million without jobs. But in 1939, the recovery resumed, partly because of numerous government contracts for military materials needed by countries that were already at war. The Spanish Civil War (1936–1939) was in progress, the Japanese had invaded China (1937), and the Soviet Union, France and Great Britain were fulfilling pledges by supplying war materials to Poland, Albania, Romania, and Greece, already being threatened by Hitler's Germany (1938). Europe was a powder keg about to blow up into World War II.

In 1939, industrial designer Harold Van Doren designed a refrigerator for Philco, a Philadelphia company just renamed from its former title of the Philadelphia Storage Battery Company. Philco had produced radios and batteries for twenty years, but this was its first refrigerator. It had earned industry respect by pioneering the use of foam in refrigerator insulation and in 1938, by introducing the first hermetically sealed window air conditioner under the brand name Philco-York, featuring a beautiful wood front and needing only to be plugged into an electrical outlet.

Mechanical air conditioners with refrigerated air were still very expensive, but there was a much less expensive option available, called an evaporative cooler or air cooler. Such coolers had no compressors, but only a fan, blowing air over evaporating water to cool it and send it into a house. They were developed in the Southwest (Arizona and California), because they were particularly effective in dry climates, since they added humidity to the cooled air. Some, called indirect coolers, passed air over a water-cooled coil; others, called direct coolers, cooled by direct contact of air with water. Early models had developed in the early 1900s, consisted of wooden frames with wet burlap cloth with fans forcing the air into the space being cooled. Later, they evolved into a configuration of two-inch excelsior pads of aspen wood shavings sandwiched between chicken wire and nailed to the cooler frame, and began to be called swamp coolers, desert coolers or wet boxes. The first of such a type was demonstrated in the Adams Hotel in downtown Phoenix in 1916.

In 1939, Martin and Paul Thornburg, professors at the University of Arizona, published mimeographed instructions: "Cooling for the Arizona Home." They had conducted tests to improve the performance of this type of cooler. About that time, the Emerson Electric Company began the manufacture of evaporative coolers, and others were mass-produced by Goettle Brothers, Inc. Soon, many houses and businesses in the Southwest would be using swamp coolers, and by the early 1950s, a large number of manufacturers would be producing such coolers for much of the United States and Canada, as well as exporting them to Australia. The appeal of swamp coolers was partly due to their relative lower cost compared to true refrigerated air conditioners, but also because of the fact that they added welcome humidity to the dry air, an important comfort issue. The disadvantage was that the re-circulating water could collect in the sump, and if not regularly cleaned, could become stagnant and foster mold and algae, thus becoming unhealthy and emitting a stale odor.[18]

Also in 1939, the Packard Motor Car Company offered the first auto air conditioner as an option for $274 ($4,600 in today's dollars) in one of its models. The equipment, called a Weather Conditioner, was made by the Bishop and Babcock Company of Cleveland, Ohio, and included a heater. The main evaporator and blower took up half the trunk space. The compressor ran off the engine and the system had no thermostat or shut-off mechanism other than switching the blower off. It discharged cool air from the back of the car, but it was awkward to operate, because the owner had to stop the car and open the hood to turn it on and off. The option was discontinued in 1941. Cadillac would experiment with air conditioners in 1941, as well.

In 1939, General Electric introduced a refrigerator with one section for frozen food and ice cubes, and another for chilled foods. It produced 480 large ice cubes in 24 hours. Kelvinator debuted a similar concept with the first across-the-top freezer in a refrigerator. Sears sold its one millionth Coldspot refrigerator at the same price as any other model: $99.50 ($1,

670 in today's dollars). Westinghouse introduced a Meat Keeper to keep meat fresh for a whole week. By 1939, Raymond Loewy had left Sears and was working for Frigidaire, also on refrigerators. He designed Frigidaire refrigerators in 1939 and 1940, on which he repeated some of his most successful Coldspot design concepts. Refrigerator design by this time had evolved into a standard configuration that would be typical for many years in the future.

Frigidaire by Raymond Loewy, ca. 1940.

On April 30, 1939, the future appeared in the form of the New York World's Fair in Flushing Meadows, officially titled "The World of Tomorrow." The Depression was subsiding, and people were not only seeing the end of the tunnel, but were anxious to move into a future filled with modern conveniences, technological wonders, and jobs for all. In short, the fair promoted the American Dream. It envisioned such a world with its dominant modern architectural symbols, the Trylon and Perisphere, and its guidebook proclaimed: "The true poets of the twentieth century are the designers, architects, and the engineers who glimpse some inner vision, create some beautiful figment of the imagination, and the translate it into a valid actuality for the world to enjoy."[19]

Industrial designers, some of whom were now famous, designed many of the dramatic futuristic exhibits and buildings at the fair, including scale models of interstate highways, new cities serviced by above-ground freeways, and worker communities in green, rural settings. New streamlined trains were introduced and demonstrated in speed runs across the country, and most major corporations had exhibits of their latest products. Walter Dorwin Teague was not only one of the eight directors of the fair, but was also the designer of exhibits and buildings for the National Cash Register Company, Ford, Kodak, and U.S. Steel. He designed the DuPont building, the pavilion of the U.S. government, and the City of Light Diorama for the Consolidated Edison Building. He also designed the House of the Future, which featured a gas-powered refrigerator designed for Norge.

There were other refrigeration exhibits there. The Westinghouse exhibit in Production and Distribution Zone 5 included all-electric modern kitchens including refrigerators, and General Electric's exhibit hall included a model electric appliance store with the full line of GE products. The Carrier Corporation pavilion was shaped like a giant igloo (actually, more like a white rocket nose-cone) with "snow" encrusted on walls and the "Northern Lights" glowing from the ceiling. Carrier displayed and explained air-conditioning systems. A large revolving globe showed visitors how air conditioning was used all over the world, from the Arctic Circle to the Equator. Although Willis Carrier's igloo at the fair gave visitors a glimpse into the future of air conditioning, domestic air conditioning would not become popular until after the war.

Other major industrial designers such as Henry Dreyfuss, Raymond Loewy, John Vassos, Norman Bel Geddes, Egmont Arens, Gilbert Rohde, and Russel Wright all designed exhibits at the fair. The Radio Corporation of America (RCA) introduced the first public television transmission with a televised speech by President Roosevelt and exhibited the first RCA television sets, designed by industrial designer John Vassos. But in September 1939, a month before the fair closed, World War II officially began with the invasion of Poland by Nazi Germany. The fair reopened in April 1940 but was significantly changed, reflecting the clouds of war swirling in Europe. The new theme for the fair was "For Peace and Freedom." Internationalism was replaced with patriotic nationalism. Ten nations withdrew due to their countries being occupied by Germans. By the time the fair closed in October, it had become a sort of county fair, although the major exhibits were still open.

In 1940 there were four major manufacturers of refrigerators: General Electric, Westinghouse, Kelvinator (owned by American Motors), and Frigidaire (owned by General Motors). That year, industrial designer Norman Bel Geddes designed a refrigerator for Kelvinator, and industrial designer Raymond Loewy designed the L300 refrigerator for Electrolux, which had a streamlined, modern look and a spring-assisted door that opened at a touch. It became a huge sales success. Westinghouse debuted a Colder Cold room air conditioner that, with a "reverse cycle," cooled or heated a room. Hotpoint, a division of GE, introduced the forerunner of the popular side-by-side refrigerator, but it was nearly twice the size of a standard refrigerator, was far too wide for most kitchens, and far too expensive for most pocketbooks.[20] Crosley, a pioneer in radio manufacturing in the 1920s, released a refrigerator with a built-in radio on the door in 1940. Crosley had put a hanging wire rack for eggs inside the door of one of its refrigerators in 1939.

By this time, the Depression was virtually over. A housing boom was in progress, and people were spending again. Most households—80 percent—were wired for electricity, and 52 percent had refrigerators. But now, the problem was inventory shortages of household goods, including refrigerators, and many stores were limiting their sales to regular customers. Manufacturers could not keep up with increasing customer demand, and were in the process of preparing for huge increases in production of cars and appliances, when in May, the government began to offer lucrative defense contracts for essential military supplies and equipment. Although America was still neutral, it was now providing tons of military equipment and war materials to Britain and France, since they were now at war with the axis powers of Germany and Italy. Consumers of appliances would have to wait.

Millions of new jobs were provided by these government contracts, and unemployment fell to 2 percent. They were well-paying jobs, and consumers had money to spend, but production of domestic products still failed to keep up with demand. By December 1941, when the Japanese attacked Pearl Harbor and the United States declared war on Germany and Japan, the country went on a full-time wartime footing, and all manufacturing companies were required to devote all their production resources to the war effort. There would be no domestic production of cars or appliances until after the war.

During the war, General Electric made gun turrets, howitzers, bazookas, and aircraft engines. Servel made wings for the P-47 Thunderbolt warplane. Norge made bullets, gun turrets and airplane parts. Frigidaire made Browning .50-caliber machine guns and aircraft propellers. Westinghouse made plastic helmet liners for the army. Kelvinator made military supplies. Coolerator made storage units, ammunition containers, mess tables, and large refrigeration units for the army. Gibson manufactured 1,078 Waco CG-4 troop and cargo assault gliders.

Some commercial chillers were removed from department stores for installation in military production plants. They would be returned after the war. Thousands of walk-in coolers were produced for use by the navy to keep food and other perishables fresh. Special portable air conditioners

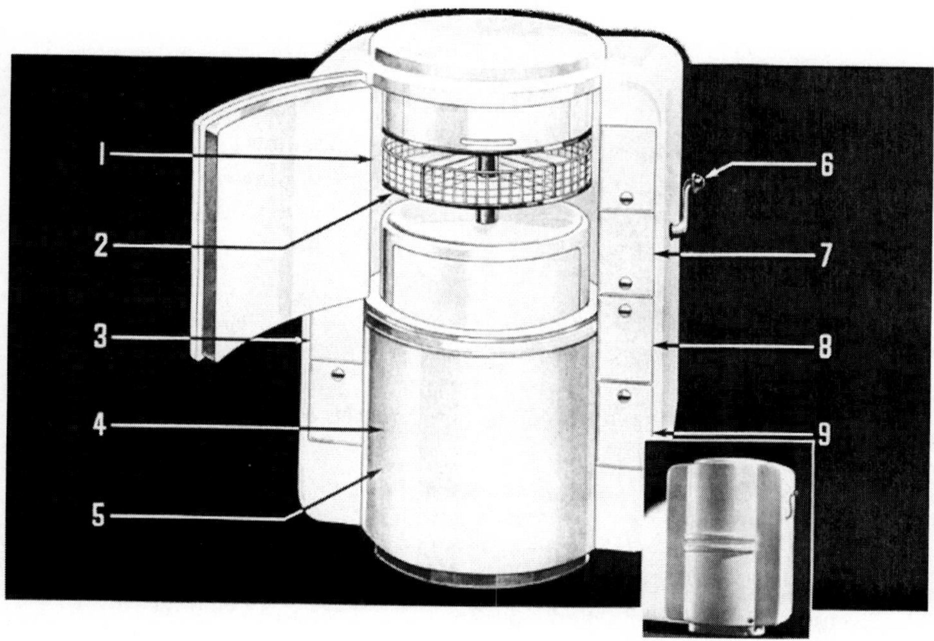

Egmont Arens refrigerator concept for Durez Plastics and Chemicals, Inc., Book Number one, 1942. (1) Upper half for general refrigeration, (2) revolving shelves make contents easy to reach, (3) cooling locker drawers for tall bottles, (4) lower half for frozen foods—kept at 10° F. or lower, (5) violet-ray compartment for sterilizing and tenderizing meats, (6) ice cube ejector lever, (7) ice cubes drop into this drawer for easy removal, (8) cold water faucet inside door, (9) cooling locker (courtesy Victoria Matranga).

were made for servicing airplanes in hot climates. Willis Carrier designed a system for the National Advisory Committee for Aeronautics to simulate freezing, for the testing of airplanes in high-altitude conditions, an environment that up to that time was thought by most to be impossible for aviation.

Many industrial designers not in military service were busy during the war designing postwar concepts. In 1942, Dow Chemical Company, manufacturer of Durez plastics, had a number of well-known industrial designers design consumer product concepts of the future for its magazine advertisements. Egmont Arens chose to design a refrigerator that attempted to solve the traditional problem of user access to items in the back of the refrigerator by making the entire food storage compartment a rotating cylinder, with pull-out storage drawers beside it. It was similar in concept to the early 1930s Westinghouse concept models that included rotating cylinders designed by Donald Dohner. The design, like many futuristic concepts, was never produced, but would inspire later 1956 GE designs with rotating shelves on a central pole inside the traditional rectangular box.

In 1944, with the end of the war in sight, the Libby-Owens-Ford glass company constructed the first of a series of many postwar, experimental "Kitchens of Tomorrow"; it was exhibited at Macy's in New York. Designed by industrial designer H. Creston Doner (1902–1991), the kitchen encouraged the use of glass in cabinet doors and had a prototype refrigerator accessible from both sides, placed between the kitchen and dining room. Visitors were asked to vote for features they wanted most to become available after the war. Carrier was planning a window-type room cooler, designed by industrial designer Henry Dreyfuss to be introduced in 1945 (U.S. Patent D144,452).

With the war over in August 1945, manufacturers rushed to convert wartime production into consumer products. The economy exploded with demand that had been dormant for 15 years, and many consumers had accumulated considerable savings from well-paying jobs during the war. Average weekly pay had doubled from $24 in 1940 to $44 in 1945. Millions of returning servicemen and -women needed cars, homes, furniture, and appliances, and they wanted the latest styles. The "World of Tomorrow" as advertised at the 1939 World's Fair was now at hand. It took about six months, but soon most major appliance manufacturers were back up to high-volume production, filling the insatiable public demand for more of everything imaginable. For the first time, refrigerators were being truly mass-produced.

Refrigerators were at the top of the consumer demand list, because only 50 percent of households had one before the war started in 1941. One of the last prewar refrigerators was the 1941 Westinghouse model designed by Ralph Kruck, then head of the industrial design department. No new refrigerators were available during the war, of course, so the other 50 percent of households were still using iceboxes and getting ice delivered to their homes by icemen. In 1946, 30,000 room air conditioners were produced, so that people who had only experienced air conditioning in movie theaters now had that luxury available for their homes. In 1948, 74,000 were produced.

One might well wonder why, 18 years after affordable household mechanical refrigerators were first introduced to the market in 1927, half of the households did not own one and still used ice in iceboxes. The simple answer is that buying ice was much cheaper, in the short term, than buying a mechanical refrigerator, and there was no 24-hour drain on the electric bill. But while the ice industry had lost half of its business for iceboxes since the 1930s, it had also expanded its business into the growing air-conditioning field. This began in the late 1920s, when the ice industry had felt threatened by the surge of mechanical refrigerator sales and searched for a means to replace its lost business. One possibility was to use ice for comfort cooling for those same customers. Beginning about 1929, numerous articles began to appear that promoted comfort cooling using ice. This possibility was discussed at President Herbert Hoover's 1932 Conference on Home Building. The conference concluded that mechanical equipment was still too expensive to make home cooling feasible, and that for a house costing $10,000, the installation of such equipment should not, ideally, exceed $5,000, half the cost of the home itself.

Westinghouse refrigerator, 1941, designed by Ralph Kruck (courtesy Hampton Wayt).

The ice industry recognized this as a golden opportunity, because equipment for cooling by ice was, on the short term, well within that maximum number. The National Association of Ice Industries (NAII) decided to do something about it. NAII concluded that the potential for comfort cooling was so vast that the ice industry would have to double its production to meet expected demand. Equipment sales were seen only as a means to increase the use of ice. NAII produced engineering manuals that, in comparing ice and mechanical systems, manipulated the calculations to show that the expense of ice cooling systems was much less than mechanical systems. Their strategy apparently met with some success. Articles in

Ice and Refrigeration magazine reported that ice was being successfully used in cooling a number of theaters, and that doubling the price of admission compensated for the daily cost of ice. The ice industry did not directly manufacture or install air-cooling equipment, but other companies were quite willing to do so.

A typical ice installation in 1937 was for the offices of the Knickerbocker Ice Company on the 21st floor of the Leggett Building in Detroit. The ice company used the system, engineered by the Typhoon Air Conditioning Company, as a showcase to promote the use of ice for cooling. The air was passed over 1,100 square feet of surface cooled with chilled water, which was chilled by spraying it over cakes of ice in an insulated bunker holding six tons. It could reduce an outside temperature of 95°F to an inside temperature of 83°F. The ice supply was recharged when 80 percent to 95 percent of it had melted. Due to ice cooling promotion, a number of comfort cooling systems installed in the 1930s did use ice. But on the other hand, mechanical cooling equipment improved and was reduced in cost, and by the end of World War II the use of ice for cooling began to decline rapidly. At the same time, with the increasing availability and reduced cost of mechanical refrigerators, iceboxes and the sale of ice for them declined rapidly as well.[21]

The natural ice industry was still strong, however, and when Eugene O'Neill's book and Broadway play *The Iceman Cometh* first appeared in 1946, everyone still knew very well what an iceman was. O'Neill wrote the play in 1939. It is set in a 1912 "Last Chance Saloon" and rooming house in Greenwich Village. Drunken roomers are contemplating their lost faith and dreams when hardware salesman Theodore Hickman ("Hickey"), played by James Barton, arrives and tries to convince everyone they can find peace of mind by bringing them back to reality. As their dreams are destroyed one by one by Hickey, they turn on him in anger with nasty accusations about his wife and the iceman. That's the only reference to the title of the play, and it didn't reflect well on icemen.

O'Neill's script continued to give icemen a bad name for the rest of the century and beyond, long after there were no more icemen. O'Neill's play would return Off Broadway in 1956, starring Jason Robards as Hickey, and he repeated his performance for a 1960 television production. In a 1973 film version, Lee Marvin played Hickey, and that same year, James Earl Jones played Hickey on Broadway. In 1985, Robards again played Hickey on Broadway, and *Iceman* was revived in 1999 with Kevin Spacey as Hickey, again in 2012 starring Nathan Lane, and as of this writing will again be revived in 2015. It seems that icemen never die! But of course, the ice trade was by the end of World War II pretty much on the ropes, due to the availability and low cost of mechanical refrigerators. When people needed ice cubes, they got them from their own refrigerators, or, if they needed larger quantities, they could buy them by the bag at supermarkets. Most major food-related businesses, restaurants, hotels and motels had their own ice-making machines. Soon the iceman would go the same way as the coal deliveryman, when electric and oil-burning furnaces replaced coal.

The refrigeration industry, of course, was up and running at full speed after the war. In 1945, Westye F. Bakke founded the Sub-Zero Freezer Company in Madison, Wisconsin. According to official company history, he invented the first freestanding freezer in 1943. Up to this time, all domestic refrigeration products had been made in white, but color was now making its first appearance in the 1947 Ben Hur home and farm chest-type freezer, made by the Ben Hur Manufacturing Company of Milwaukee. It was blue (as are some arctic icebergs), and was designed by Milwaukee industrial designer Brooks Stevens (1911–1995). In 1947, Amana made the first upright freezer for the home (an honor also claimed by Gibson).

Why this sudden popularity of home freezing capacity? The often-small initial freezer compartments in refrigerators before the war, essentially for only ice cubes, could not accommodate the rapid growth of frozen foods, which had become highly popular during the war. Busy families with wartime jobs found it inconvenient to cook dinners from scratch, finding it much easier to thaw frozen foods that were not only ready to eat, but tasted just like fresh food. The popularity of frozen food was attributable to just one man, the founder of the frozen food industry, Clarence Birdseye (1886–1956).

Frigidaire had produced a retail combination display and storage cabinet in 1929 for Birdseye, the same year that Birdseye sold his company and patents for $22 million to the Postum Cereal Company, which became General Foods Corporation that year and founded the Birds Eye Frozen Foods Company. Conventional freezing methods in the 1920s commonly used higher temperatures, and freezing occurred relatively slowly, giving ice crystals time to grow and causing damage to the tissue structure. When the frozen food thawed, cellular fluid leaked from the damaged tissue and gave the food a mushy consistency that was unappetizing and dry. Birdseye solved this problem.

From 1912 to 1915, Birdseye, then a taxidermist, was assigned to Labrador in Newfoundland, where the Inuit American Indians taught him how to ice fish in –40°F weather. He noticed that the fish he caught would freeze almost instantly, and when thawed, tasted fresh. He realized the potential in an industry where frozen food was normally of a low quality because of the higher freezing temperature. In 1922 he conducted experiments, and by 1925, had developed a flash-freezing process and formed the General Seafood Corporation.

His process froze fresh fillets at −45°F using his invention, the patented double "belt froster," in which cold brine chilled a pair of stainless-steel belts carrying packaged fish between them, freezing the fish quickly. This equipment was patented in 1930 (U.S. Patent 1,773,081). In 1927, he extended the process beyond fish to meat, poultry, and vegetables. He still couldn't sell the product and decided he needed a partner in the food industry that could provide effective distribution. General Foods, a cereal manufacturer, agreed to be a partner in 1929, and began by offering free samples in shops and by having speakers give talks to local groups. They diversified into frozen fruit juices. Business picked up. Birdseye sold his existing rights to General Foods, which included using his name, but it was split in two to "Birds Eye" and was registered as a trademark in 1931. That year he went to Frigidaire, requesting special-purpose

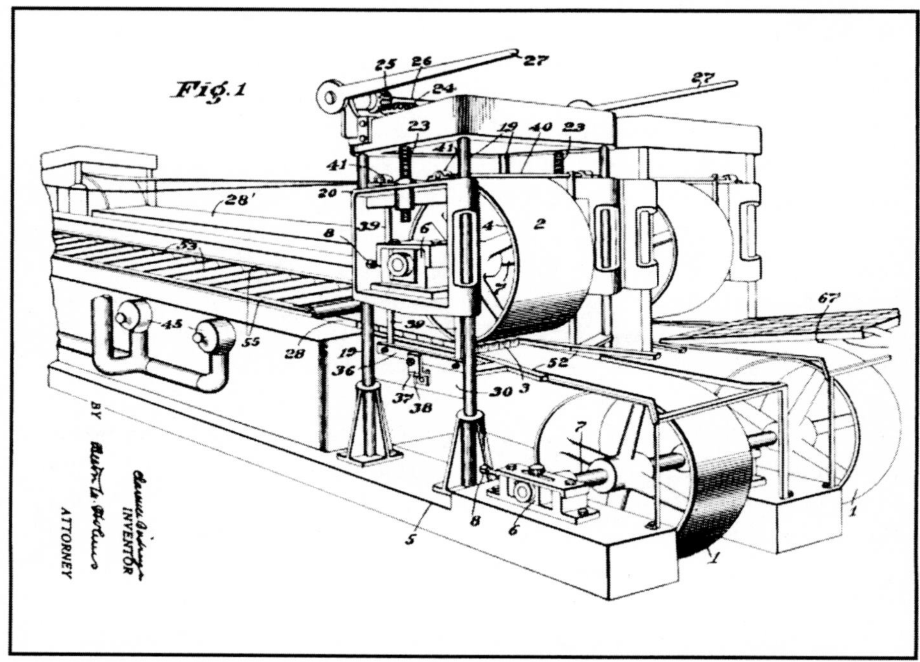

Birdseye "belt froster," 1930. U.S. Patent No. 1,773,081.

refrigeration, which resulted in the retail storage and display cabinet for Birds Eye frozen food. In 1930, the company began sales experiments in 18 stores in Massachusetts, which tested a product line of 26 items, including 18 cuts of frozen meat, spinach, and peas, a variety of fruits and berries, Blue Point oysters, and fish fillets. Customers liked the products, and popularity of frozen food grew during the 1930s, when many people had refrigerators with freezing compartments, although they were small.[22]

Frozen food received a political boost during 1936 to 1938, when Joseph Edward Davies (1876–1958) was the second ambassador to the Soviet Union, appointed by President Franklin Roosevelt. Davies' wife was Marjorie Merriweather Post (1887–1973), the daughter of C.W. Post, the owner of the Postum Cereal Company that became General Foods in 1929, which had bought the rights to Birds Eye that year. Marjorie Post was in 1936 the wealthiest woman in the United States, with a fortune of about $250 million ($4.2 billion in today's dollars). She deployed commercial grade freezers to Spaso House, the U.S. embassy in Moscow, in advance of their arrival, because she was fearful of the Soviet Union's food processing standards. She stocked the freezers with products from General Food's Birds Eye unit. The frozen stores allowed the Davies to entertain lavishly and serve fresh-frozen foods that would otherwise be out of season. Upon returning from Moscow, Post (who would resume her maiden name after divorcing Davies in 1955), directed General Foods to market frozen products to upscale restaurants, increasing its prestige.

During the time Birdseye was developing flash freezing, another type of freezing, called freeze-drying, was being developed by the Nestle Company for the preservation of coffee, after the company was asked by Brazil to help find a solution to its coffee surpluses. Nestle produced their own freeze-dried product, called Nescafe, which they introduced in Switzerland in 1938. The process, technically called lyophilization, essentially freezes a product and then removes all moisture at low temperature under a vacuum, causing the frozen moisture to vaporize without passing through a liquid phase. The process, which retains the natural flavor and color of food, with no need for cooking or refrigeration, was originally used by the ancient Peruvian Incas of the Andes. They stored their potato and other food crops on the mountain heights above Machu Picchu. The cold mountain temperatures froze the food and the water inside slowly evaporated under the low air pressure of the Andes. In 1905, a chemical pump was used to produce a vacuum, and scientists realized that the material had to be frozen before commencing the drying process. During World War II, the freeze-drying process was used to preserve blood plasma and penicillin, and John Glenn's orbiting the earth in 1962 began the development of freeze-dried foods for space travel. After the 1960s, over 400 types of freeze-dried products would be commercially produced. In the taxidermy field, freeze-drying became a preferred technique to preserve pets as realistically as possible, although the process is expensive and specimens need to spend six months in the freeze dryer.

During World War II, with many two-income families working full time, high-quality frozen food was a welcome convenience. Now, after the war, because of the rapidly increasing popularity of frozen foods, the demand for larger freezers in refrigerators and for chest-type freezers was enormous. In 1947, Kelvinator made the first two-door refrigerator with two separate cold zones—one for frozen food, the other a cooler—and General Electric made the first two-door refrigerator/freezer at its Erie plant that kept frozen food at 0°F to10°F in the freezer, and at 38°F in the refrigerator. That same year, bright red Coca-Cola coolers appeared in drug stores and lunch counters across the country, such as the one designed by industrial designer Raymond Loewy and made by Dole Deluxe. Westinghouse industrial designer Ralph Kruck designed the thousands of ubiquitous Coca-Cola coolers that appeared across the country in gas stations, grocery stores, and convenience stores. Also in 1947, mass-produced, low-cost, window air conditioners became possible with innovations by engineer Henry Galson, who set up production lines for a number of manufacturers. That year, 43,000 window air conditioners were sold, and in 1948, 74,000 air-conditioning units would be sold.

In May 1947, some members of the English Institution of Heating and Ventilating Engineers and of the Society of Instrument Technology, who met informally, decided to form what they called the Rumsford Club, named after Benjamin Thompson, Count of Rumsford (1753–1814), who was a pioneer of heating science, ventilation, and fireplace designs. Thompson was born in the American colonies (Massachusetts) but had collaborated with the British in Boston in 1774, and had to leave the country for Europe, where he spent the rest of his life. Thompson had expanded the 1793 research about smoky chimneys and chimney flues conducted by another American, Benjamin Franklin, in Thompson's publications of 1796 and 1798, which initiated his work in stove design, and his work became the basis of stove design throughout the rest of the 19th century. Thompson later was created a count for his services in military and civil fields, and became a Fellow in the British Royal Society. In June of 1947, the newly formed Rumsford Club convened on short notice in honor of Willis Carrier, who was visiting the country for the centenary celebrations

Coca Cola cooler, ca. 1941, designed by Ralph Kruck of Westinghouse (courtesy Hampton Wayt).

of the Institution of Mechanical Engineers. Carrier attended the meeting and was made an honorary member of the club in recognition of his contributions to air-conditioning development.[23] Carrier has often been called the "Father of Air Conditioning," the term apparently first used in 1949. But Carrier himself was not inclined to agree. He himself admitted in 1929: "No individual or no firm can take credit for all or any part of these developments." And although Carrier did not invent air conditioning, he was clearly its greatest advocate and practitioner in the last 90 years.[24]

Chapter 8

Postwar Refrigeration

In 1947, the Fedders Company began manufacturing room air conditioners, which they called "aircons." Founded in 1896 by Theodore C. Fedders as a producer of milk cans, bread pans, and kerosene tanks, Fedders manufactured radiators for 30 car manufacturers in the early part of the century. In 1928 Fedders had begun to make electric water coolers, refrigerator components, and "aircon" components.[1] In 1948, Frigidaire made the first refrigerator/freezer combo with a completely separate freezer section with a fan. That same year, new domestic manufacturers entered the field, when industrial designer Brooks Stevens designed a refrigerator for the Coolerator Company, which had begun to manufacture refrigerators and freezers in 1945, after it had been making large refrigeration units for the military during the war. Coolerator had been founded in 1908 in Duluth, Minnesota, as a maker of iceboxes, and in 1932 had changed its name to the Duluth Refrigerator Company to reflect its manufacture of mechanical refrigerators. By 1950, Coolerator was also making air conditioners, but by 1954, it was out of business, due to intense competition from the larger manufacturers.

In 1949, Amana introduced a side-by-side refrigerator/freezer, and Raymond Loewy designed a complete line of major appliances for Frigidaire, including refrigerators. That same year, General Electric installed ten experimental models of the first heat pumps around the country.

The General Electric experimental models in 1949 were those that are generally referred to as heat pumps to avoid confusion with the refrigeration type of heat pumps, because they produced heat, not cooling.

In the 1940s, refrigeration units were designed for trucks, which, along with railroads, would soon convey refrigerated goods around the country. Black history references identify a black inventor, Frederick McKinley Jones (1893–1961), who was credited with the invention of the first practical and automatic refrigeration unit for trucks (U.S. Patent

D132,182, granted in 1942 and assigned to the U.S. Thermal Control Company), a design that was later adapted to other carriers, including ships and railway cars. He was elected to membership in the American Society of Refrigeration Engineers (ASRE). The trucks would use the new interstate highway system, planned since 1939, which would be constructed starting in 1956, authorized by the Federal Highway Act under President Eisenhower.

Coolerator refrigerator, 1948, designed by Brooks Stevens.

There has always been some confusion in the mind of the general public between an "invention" and a "design." Since 1790, when Congress established the U.S. Patent Office, patents were issued (initially for 14 years, but later, 17) to inventors for "any useful art, manufacture, engine, machine or device, or any improvement thereon not before known or used." Historically, patents were regarded generally as mechanical inventions. In 1842, the Patent Office convinced Congress to pass the "Act to Promote the Progress of the Useful Arts," which granted patents for "any new and original shape or configuration of any article of manufacture not known or used by others before." The intention clearly was to protect the unique appearance of objects, rather than their mechanical or operational aspects. These were called "Design Patents," and were numbered chronologically with a "D" preceding the patent number, in a different numerical series than mechanical patents. They were legal recognition that appearance invention was just as important as mechanical invention.

Fast-forward to the late 1920s and 1930s, when industrial designers entered the manufacturing scene and began to design the external appearance of almost all manufactured functional products. Their designs were, and still are, protected by Design Patents, issued by the Patent Office, and have a "D" before the patent number. So when one wants to distinguish conversationally between mechanical patents and design patents, despite the fact that they are both inventions, one of mechanical and the other of appearance, the term "invention" usually refers to a mechanical or operational invention, and the term "design" usually refers just to the configuration and appearance of a device. It is not a minor distinction, since the term invention or inventor, to the general public, conjures up people like Thomas Edison or Steve Jobs, world famous inventors. When someone knowingly or unknowingly calls a designer an inventor, it elevates the contribution of that individual to a higher status in the minds of the general public, who are often unaware of the distinction between the technical types of inventors. So to clarify the reference to Frederick Jones above, he was actually a designer, not an inventor, and his patent number, since a "D" precedes it, was actually an external appearance design or configuration, not an invention in the mechanical sense. This is not to belittle his contribution, but to more accurately clarify its nature. Since the refrigeration device he designed was to appear mounted on the exterior of trucks, special attention needed to be paid to how it looked. He did that quite effectively.

By the 1950s, most Americans had electric refrigerators. Refrigerators, once a stand-alone appliance, were now only a single element in an integrated, modern kitchen environment. Manufacturers had to innovate new mechanical and appearance features to attract consumers. This chapter describes some of these and how increased competition forced the consolidation of manufacturers.

A major study in 1950 showed that families living in air-conditioned homes slept longer in summer, had more leisure time, and enjoyed food more than families without it. Air conditioning for commercial buildings allowed skyscrapers to be built and freed architects to build how and where they wanted, which led to a building boom. Air conditioning transformed residential construction in the South, as well, when ranch-style homes replaced Victorian-style homes with walk-around porches to harvest natural breezes. Ceilings were lowered from ten to twelve feet or more to eight. Windows were reduced in size and sleeping porches were eliminated or converted into sunrooms.

Also in 1950, at a time when 90 percent of American homes had refrigerators, industrial designer John M. Little (1906–1996) started design work on a new line of refrigerators and freezers made by the Bendix Home Appliance Division of AVCO Manufacturing Corporation. It was the company's initial venture into the kitchen white-goods field. The Bendix name in home appliances had been well established in the washing machine market after the war, with its famous twelve-inch, round porthole on the front door so that the washing action could be seen. It was not mere coincidence that watching the action was similar to watching the television screens now becoming popular in American homes. The dominant design feature of the Bendix refrigerator was its six-inch-diameter acrylic ring that served as the latch and grip for opening the door. The circular handle and trademark motif suggestively echoed the popular and familiar round porthole of the Bendix washing machine, by now a brand identity.

That same year, Westinghouse debuted its Frost Free refrigerator, the first with automatic defrost in both the refrigerator and freezer compartments. This feature ended the weekly and annoying routine of manually defrosting the build-up of frost, which if ignored made the removal of ice-cube trays virtually impossible. Other manufacturers would soon follow suit. In 1951, Philco would introduce its two-door refrigerator with automatic defrosting, and in 1952, Frigidaire would debut its competitive automatic defrosting feature, Cycla-Matic. The Fedders Company claims that in 1934, it perfected the non-frosting systems for its commercial refrigerators with its frost-free coil technology, which preceded by 20 years the introduction of frost-free refrigerators to the domestic market.[2]

The Korean War began in 1950, and the U.S. defense program restricted the use of brass, chrome, and plastics on consumer products, including refrigerators. For this reason, industrial designer Peter Muller-Munk (1907–1967) had to remove a decorative chrome plate behind the handle latch of the 1951 Westinghouse refrigerator he had designed.[3] He, and other designers, actually welcomed these restrictions, because they reduced the amount of cosmetic and garish decorations often demanded by company sales or marketing personnel. Removal often meant cleaner designs, which were preferred esthetically by the Museum of Modern Art's (MoMA) "Good Design" program that ran from 1950 to 1955. MoMA, dominated by architects, collaborated with the Merchandise Mart in Chicago, hoping to convert conservative manufacturers of furniture to modern design. Each year, MoMA mounted "Good Design" exhibitions, declaring 250 or so products selected by the museum from the Mart as "Good Designs" with a label declaring: "The manufacturer guarantees that this article corresponds in every particular to the one chosen by the Museum of Modern Art, N.Y. for the Good Design exhibition at the Merchandise Mart, Chicago."

Products selected and attached with MoMA "Good Design" labels were promoted nationally, presumably as a desirable sales feature. MoMA's criteria for selection were primarily esthetic in nature, preferring clean, simple, geometric forms, with absolutely no decorative trim, and with a brand name that was discrete and simple in typeface. Streamlining and bold decorative features were specifically rejected. MoMA's attempt to control the style of product design, in order to educate the public on "good design" and to control the industrial design profession, was generally unsuccessful with the public. Prominent industrial designers were not impressed, either. They regarded such esthetic criteria as inadequate to appreciate the fundamental qualities of truly good design, which, in addition to appearance, included ergonomics, consumer satisfaction with operation, manufacturing ease, affordable consumer cost, and success in the marketplace. Henry Dreyfuss, when asked by visitors from an English museum, which museums in which he would like to have his designs exhibited, he replied: "They are called Macy's, Marshall Field's, and the May Company—similar to your Eaton's, Simpson's, and Morgan's."[4]

Department stores, of course, are where sales are made, and industrial designers gauged their success by sales, not museum recognition. Another prominent designer went further, by stating that to have one's design chosen by a museum was the "kiss of death" in the marketplace. Nevertheless, MoMA would establish esthetic standards in the minds of many designers. As one of the industrial design associations began to make the first annual, national design awards in 1951, esthetics became the major criteria for such awards, ignoring all merchandising realities. Basically, they are beauty awards, not unlike the Miss America Pageant. What you see is what matters. Most refrigerator manufacturers relied not just on appearance, but on features of convenience for users. In 1952, Crosley introduced a chilled drink dispenser, a rudimentary tank that allowed the user to store anything from juice to milk to water. Turn a spigot, and the beverage flowed through the door.

By the time of the Korean War, the natural ice industry begun by Frederick Tudor in 1806 had changed dramatically. Because about 90 percent of all American homes now owned electric refrigerators, only a few rural areas continued to use manufactured block ice in household iceboxes. By the 1960s few block-ice plants would remain and by the 21st century, there would be only about 50 such plants nationwide, mostly in the South, that would find customers in produce, ice sculpture, and the movie industry. To most people after the 1950s, "ice" meant mechanically manufactured and packaged ice cubes, the kind you buy in bags at grocery or convenience stores to fill beer or soda-pop coolers for picnics or holiday parties. These cubes were usually produced by locally operated ice plants.

For the domestic refrigerator industry, there were also problems. The market was saturated and most sales were for replacements of older units. Refrigerators lasted from 15 to 20 years, a long time compared to other major appliances. Competition had become intense in both refrigerator design and in consumer pricing. Servel, Inc., was in serious financial trouble when its sales plummeted from $71 million in 1948 to $41 million in 1949. The problem was household refrigerators, which accounted for 70 to 80 percent of the company's dollar volume. The fact that the refrigerator market was 90 percent saturated was a problem for all refrigerator manufacturers, not just Servel. It was a time where many manufacturers introduced new mechanical and operational features, as well as new colors and innovations, to tempt customers to replace their old refrigerators with new ones. Styling by industrial designers was more important than ever.

But Servel had an additional, more serious problem: the U.S. Justice Department had slapped it with an antitrust suit, which ended its exclusive distribution arrangements with gas utilities. Servel had won its name and fame as the manufacturer of absorption-type refrigerators run by gas (and also by kerosene). With no moving parts required to reclaim the refrigerant in absorption models (motors do the job in electric compression models), Servel had long used the sales pitch: "stays silent, lasts longer." In addition, deaths from carbon monoxide poisoning from faulty gas refrigerators rose alarmingly in New York City.

Servel responded to these serious challenges by bringing in as president W. Paul Jones, formerly vice president of refrigeration at Philco. Jones decided on a diversification program that would increase revenue with new products. First, he ended the 25-year marriage to the gas industry by announcing in 1951 that Servel would make absorption refrigerators that ran on electricity, as well as gas, and that later, the line would include regular electric compression refrigerators. The new housing developments going up all over the country generally lacked gas lines but operated primarily on electricity, another deterrent to the gas refrigeration industry.

During the Korean War, in 1952, Servel began production of a $100-million contract for jet airplane wings for Republic Aviation's Thunderjet and Thunderstreak, and sales from civilian projects for the first nine months of 1952 totaled $50 million. Servel employed about 13,000 people, who were relieved that things were going well despite the setbacks. Servel's new line in 1953 would include refrigerators, air conditioners, vertical and horizontal home freezers, and water heaters.

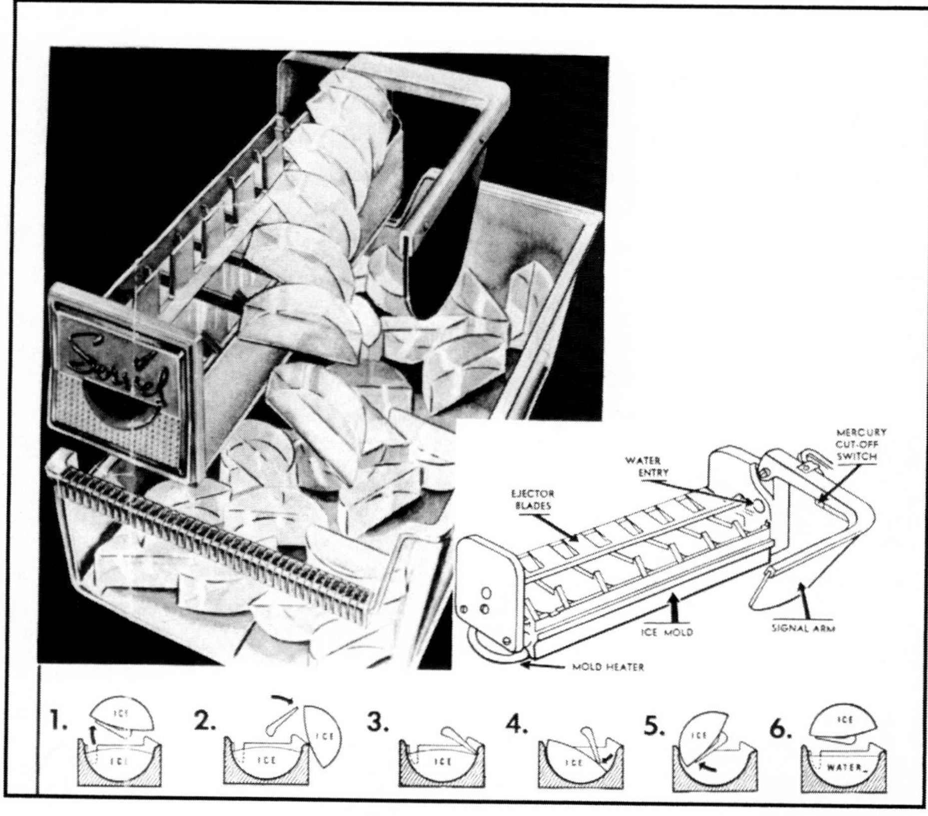

Servel ice maker, 1953.

Servel's ace in the hole was its revolutionary new Servel ice maker, under development since 1947. Without bothering with ice trays or any effort on the part of the user, the ice maker kept a large container continually filled with semicircular ice "cubes" (the kind we are accustomed to today; the semicircular shape enabled them to slip easily and automatically out of their freezing tray, after a heating element loosened them, and when the tray was tilted). The ice maker used an electric motor, an electrically operated water valve, and an electrical heating unit. To provide power to all these elements, one had to hook up the ice maker to the electrical circuit that powered the refrigerator. Also, one had to hook the ice maker up to the plumbing line in the house, to provide fresh water for the "cubes." The power line and water line both ran through a hole in the back of the freezing compartment. A Servel spokesman said that if you left the door

Servel Wonderbar, 1953. Designed by Don Dailey.

open, it would fill the whole kitchen with ice. The automatic ice maker cost about $300 to $600 (installed) depending on the size of the unit.

Servel also wanted to get the refrigerator into areas other than the kitchen. The most innovative refrigerator design of the year was the 1953 Servel Wonderbar portable refrigerette, the brainchild of vice president Donald Dailey (1914–1997), a noted industrial designer hired by Jones in 1950 from Philco as new products manager. The size of an ice chest, with no more than a two-cubic-foot capacity, the Wonderbar looked like a piece of furniture, appropriate for game rooms, finished basements, bedrooms or offices. It was an absorption-type machine, run on electricity. Another new Servel product in 1953 was the Wonderair, a room air conditioner designed by Dailey that also looked like furniture, and was controlled not by buttons or dials but by pulling out the conditioner, mounted on a drawer, to any of three positions. To launch its new 1953 line, Servel took it on nationwide tour of twenty-nine cities and featured it to dealers with a five-hour musical extravaganza that played in each city's leading theater.[5]

By 1953, one million room air conditioners had been sold in the consumer market. In 1950, annual sales had exceeded 100,000 units, and would never drop below this level again. Most were designed to blend in with window dressings and furnishings. Westinghouse's model included three square directional louvers that could be adjusted by the user to direct the air in any of four directions. Industrial designer Walter Dorwin Teague designed GE's models in "fiesta tan," fawn, or brown. Other new manufacturers were entering the field. International Harvester (IH) had begun making refrigerators, freezers and air conditioners in 1947. A traditional maker of farm tractors, IH had a refrigerator division, just like other auto manufacturers (Ford had Airtemp; General Motors had Frigidaire; Nash-Kelvinator Corporation, later American Motors, had Kelvinator; Studebaker had the Franklin Appliance Company; and Crosley Motors had Crosley refrigerators). In about 1953, the IH refrigerator line, developed by Chicago industrial designer Dave Chapman (1909–1978) and IH staff designers, included the Decorator, a model with a specially designed door with a welded-in front sheet, upon which was bonded a colorful vinyl-coated decorative fabric. All units, however, could be decorated

International Harvester refrigerator, with changeable décor, c. 1953. Designed by Dave Chapman, Inc. (courtesy the Industrial Designers Society of America,* Industrial Design in America, *1954).

with the user's choice of fabric to match any kitchen decor. It was the first major appliance whose color and texture was unlimited, and it could be easily changed to fit any decorative scheme. The interior of the IH freezer that year, also by Chapman, was in color: "Spring-Fresh Green."

Most refrigerators since the mid–1930s had doors that opened from left to right, with the latch handle on the left. This suited most right-handed users, who used their right hand to open the door and removed or replaced food containers with the left. But depending on various kitchen arrangements or left-handed users, some customers preferred doors that opened from right to left. This was possible on a special-order basis. But in 1953, Philco introduced its "V-handle" model, which had a unique door design that opened in either direction. When the left side of the "V" was pulled, the door opened from the left, and when the right side of the "V" was pulled, it opened from the right. It was a selling feature that allowed the refrigerator to be bought for one house, but also moved and used in a new house if the reverse door-opening situation was needed.

The handle was an ingenious design but very complicated. Any door needs a hinge on one side and a latch on the other. The Philco had two chrome rails on both sides of the cabinet. The door had a complicated mechanism with a series of small claws running along the perimeter of the door that acted as either the hinge or the latch, being determined by which way the big V handle was pulled. Harold Van Doren, who was the industrial design consultant retained by Philco, invented it. The 1953 models also came in bright blue and red options, another dramatic decorative feature provided by Van Doren, as color in major appliances became more popular. The V-handle Philco model would become a classic refrigerator design coveted by collectors today (it is worth $2,000 to $4,000) and would be produced until 1955. Doors opening from either side would reappear as a feature in the 1960s, only by certain companies. Today, of course, doors can be easily changed to right or left by customer choice, as the installer simply moves the hinge and handle to the desired side.[6] As kitchens and their appliances got bigger, the General Air Conditioning Corporation in 1953 took a different route by making a compact unit that housed a stove, a sink, a freezer, and a refrigerator all in one. A similar model is sold today. Mamie Eisenhower's pink dress at her husband's inauguration in 1953 sparked a craze in pastel-colored refrigerators, as well as cars.

Automotive air-conditioning returned with vengeance after the war. The 1953 Chrysler Imperial was the first production car in 12 years to actually have air conditioning, following the early prewar experiments by

Packard and Cadillac. The system was called Airtemp, the same type as had been developed and installed in the New York Chrysler Building in the 1930s. Chrysler got a jump on Cadillac, Buick, and Oldsmobile, all of whom added air conditioning as an option later in the 1953 model year. Airtemp was more sophisticated and efficient than any of the complicated competitive systems. It was simple to operate, with a single switch on the dashboard marked with low, medium, and high for the driver to select. It was capable of cooling the car from 120°F to 85°F in about two minutes, completely eliminating humidity, dust, pollen, and tobacco. Because it recirculated, rather then merely cooled, the air, it avoided the stale odor that was associated with other auto air-conditioning equipment of the time. And instead of plastic tubes mounted on the rear package shelf, as on GM or other cars, small ducts directed cool air toward the ceiling, where it filtered down around the passengers, instead of directly blowing on them, a comfort feature that modern cars have lost.[7]

In 1954, Philco purchased both the Bendix and Crosley refrigerator companies. Nash-Kelvinator became a division of American Motors when Nash and Hudson automobiles ceased production and the companies merged. This merger of automotive technologies with refrigeration resulted in the production of the Weather Eye, the first true refrigerated air conditioner for automobiles, to meet the insatiable demand for such products, particularly in the South. The Weather Eye was installed in the 1954 Nash Ambassador, the first American car to have a front-end, fully integrated heating, ventilating, and air-conditioning system (HVAC). All components were installed under the hood or in the cowl area. Most competing systems used a separate heating system and an engine-mounted compressor, driven off the crankshaft of the engine by a belt, with an evaporator in the car's trunk to deliver cold air through the rear parcel shelf and overhead vents. The Weather Eye system was compact, easily serviceable, and relatively inexpensive. It cost $345 ($3,030 in today's dollars), the least expensive of all competitors. Nash's layout became an established template and is the system that continued to be used, up to the present day.

A less effective, popular, competitive economy model was mounted under the dash and was nicknamed the "crotch cooler" by the public. In 1960, Cadillac would debut a bi-level HVAC system in its automobiles. By 1962, about 11 percent (756,781 units) of all cars sold in the United States were equipped with air conditioning. This included both factory-installed and after-market systems. It was not until 1964 that automatic temperature control appeared in cars. By 1969, over half of all domestic autos had air conditioning, not just for passenger comfort, but to increase the car's resale value. No one wanted a car without air conditioning.

Amana began the manufacture of household air conditioners in 1954. That same year, Fedders introduced the first ¾-horsepower room air conditioner operating safely on standard 115V circuits, and established a new industry standard for efficient performance.[8] Also in 1954, the Deepfreeze Division of the Motor Products Corporation debuted its first refrigerators and freezers, designed by industrial designer John Harold Walter. General Electric revived the idea of rotating, lazy–Susan type, semicircular shelves in its refrigerators. The feature was called "Food at Your Fingertips," and the shelves, which were adjustable in height even when loaded, rotated on a central, stainless-steel shaft located about two inches from the front of the inside of the cabinet, and swung out for easy access. The adjustable shelves allowed space to be efficiently used by accommodating items from milk bottles to sardine cans. Unlike the rotating food compartment of 1930s Westinghouse concept models, the GE rotating shelf models were of a full depth and capacity. Unfortunately, as bored kids pushed the button that rotated the shelves, condiment jars flew off the shelves and onto the floor. Today, deep shelves in the refrigerator door provide similarly easy access to most items in the fridge.

Westinghouse refrigerator, 1954. Designed by Peter Muller-Munk (courtesy the Industrial Designers Society of America, **Industrial Design in America,** *1954).*

Frigidaire introduced the first line of color-matched major appliances in 1954. The 1954 Westinghouse refrigerator also joined the color parade with its interior in robin's egg blue. Designed by leading industrial designer Peter Muller-Munk, the Westinghouse had the first electrically operating door-opening device for refrigerators, a pad instead of a latch handle, which, when touched with hand or elbow, opened electrically. The feature was on models with

10- and 12-cubic-foot capacities. There were two roll-out shelves, one adjustable shelf, one fixed full-width shelf, and one removable half-shelf. It contained a 56-lb. freezer, two humidifiers, a beverage keeper drawer, and in the door, tilting, removable egg racks, cheese and snack-storage compartments, and fruit bins.

Concept kitchens were beginning to be developed by a number of manufacturers to promote their brand and the latest innovations in major appliances. Modern kitchen designs featured color and form-coordinated appliances, counters, and storage cabinets that were integrated into a cohesive whole, to appear as planned built-ins. Sub-Zero Company claims to have been the first to manufacture built-in refrigerators in the 1950s. The trend toward larger kitchens began to include island counters and sinks in the center of the room. In 1954 General Electric developed a kitchen concept that was featured in *Life* magazine, which included an innovative refrigerator in the center of the kitchen. It was about the same size as a traditional refrigerator, but instead of standing upright on the floor, it was mounted horizontally at eye level above a service cabinet, easily accessible from a standing position. It looked and operated like typical wall-hung storage cabinets above working counter spaces, and like such cabinets, three doors on either side swung open, to access the refrigeration and freezing compartments. The concept was developed and prototyped by the GE industrial design staff under Arthur BecVar. Hotpoint was still a refrigerator brand name made by GE, but Jean Reinecke of Barnes and Reinecke in Chicago designed this 1950s model.

Hotpoint refrigerator ca. 1950s, designed by Jean Reinecke (courtesy Victoria Matranga).

National organizations in the heating and ventilating industry were now recognizing the growing importance of the air conditioning industry. In 1953 the Air-Conditioning and Refrigeration Institute (ARI) was formed through the merger of two related trade associations. The American Society of Heating and Ventilating Engineers (ASHVE), formed in 1894, changed its name to the American Society of Heating and Air-Conditioning Engineers (ASHAE) in 1954. In 1959, ASHAE would merge with the American Society of Refrigerating Engineers (ASRE), formed in 1904, and become the American Society of Heating, Refrigerating and Air Conditioning Engineers (ASHRAE). Despite the word "American" in the name, it is an influential international organization.

By 1955, 80 percent of American households had a refrigerator. That year, Kelvinator introduced its first side-by-side refrigerator with two doors opening from the middle, called the Fooderama. It was a gigantic model, with a surplus of drawers and compartments, many of which were designed to accommodate specific products such as frozen juice cans. Whirlpool Corporation began the manufacture of refrigerators and air conditioners with its merger with the Seeger Refrigeration Company and the RCA air conditioner business. Frigidaire introduced its Ice Ejector, a storage bin with a built-in ice-cube release. But mechanical features were not the only way to attract new customers. GE in 1955, following the lead of other competitors using color in major appliances, introduced all its major appliances, as well as matching, metal, wall storage cabinets, in a choice of rich "Mix-or-Match Colors": Turquoise Green, Petal Pink, Cadet Blue, Woodtone Brown, White or Canary Yellow. Consumers could choose their favorite kitchen color among these. Where did these colors come from?

The answer goes back to 1946, when the editors of *House and Garden* (*H&G*) magazine realized that after the drab war years there would be a natural and spontaneous need for fresh new colors in decorating. But what colors? In which room? In what materials? How would they be coordinated, shown, and sold, first to retailers and distributors, and then to consumers? The answer was the House and Garden Color Program, initiated in 1946. It was conducted in cooperation with designers, manufacturers, and retailers of home products, to determine what the current color trends were, in some cases to initiate them, and to communicate them to the public. Initially, a color palette was initiated, which accurately represented current color trends but which also included future trends. The palette was intended for use by industry, retailers, and consumers. Industry could match products to individual colors. Retailers were shown how to promote and sell home products in fresh new combinations that would appeal.

Soon, the extensive research and market studies required to determine the best selling current colors and their trend line projected into the future became more than *H&G* staff could manage. Faber Birren (1900–1988), an internationally recognized color authority, became the Color Program's research consultant, a post he would hold well into the 1980s. Birren studied at the Art Institute of Chicago and the University of Chicago, and settled in Stamford, Connecticut. He was an artist, lecturer, and author of over 30 books and 200 articles on all aspects of color and human response to it, and from his New York office, Birren also consulted with a number of manufacturers and government agencies on functional color (to improve the performance of work environments, calm inmates in institutions, or communicate hazards, etc.).

From a wealth of sales data gathered by Birren from paint, wall covering, carpeting, resilient flooring, upholstered furniture, drapery, kitchen appliance, bathroom accessory, and other product manufacturers each year, *H&G* published a full-color annual graph identifying the relative quantity of sales of its palette of colors that particular year, traced that color's sales history back to 1946, and projected the trend line of that particular color on the graph into the next several years. New colors were added as they became popular, and colors were dropped from the chart when they lost appeal. The chart was publicized annually by *H&G*, and became a handy and essential reference for designers and manufacturers alike. The slogan of the program was "not every color sells every product, but every product is sold by its color." So when a manufacturer was developing its new line of products, it could look at the latest *H&G* chart and see exactly what colors were then popular, and whether that color was trending up or down competitively. Accurate color swatches and chips were available from *H&G*, but manufacturers and designers often developed their own individual variations of specific *H&G* colors.[9]

When designer Brooks Stevens selected the first non-white color for major appliances in 1947 in his design for the Ben Hur home freezer, he probably referred to the top leading popular color on the 1946 *H&G* color chart, which was, in fact, blue. International Harvester's use of spring green for the interior of its 1953 freezer, Westinghouse's use of a robin's egg blue interior on its 1954 refrigerator, and Frigidaire's color line in 1954 were all probably based on *H&G* charts showing that pastels were very popular at that time. Westinghouse and Frigidaire were probably the first companies to use colors other than white on a refrigerator. Other refrigerator manufacturers took note of the trend toward colors. By the time GE decided to abandon white for its 1955 line of major appliances, all it had to do was look at the latest *H&G* color chart, and see that pastel pink, yellow, and turquoise were among the most popular consumer colors, and were projected to remain so for several years. The *H&G* color program made selection of colors an easy and safe merchandising decision at the time and was enormously successful.

Amana "Dutch Door" refrigerator by Jean Reinecke (courtesy Victoria Matranga).

In other refrigeration news of the 1950s, Norge also introduced the first rotary compressor, called the Rollator, into the industry, permitting refrigeration units to be smaller and quieter, and to weigh less, as well as operate more efficiently than those with the traditional reciprocating compressor. Amana introduced a refrigerator with "Dutch doors," designed by Jean Reinecke. "Dutch doors," kitchen doors with an upper half that opened independently of the lower half, were popularized by postwar movies depicting idyllic suburban country lifestyles. In a similar manner, refrigerators began using two doors instead of the traditional single door, to appeal to this style trend.

"Kitchens of the Future" were still an effective way for manufacturers to promote new products and innovations to the public. In 1954, Alexander Kostellow (1896–1954) and his wife Rowena designed a Kitchen of Tomorrow for GM's Frigidaire division that was exhibited at GM's annual Motorama. In 1957, an experimental fiberglass Monsanto House of the Future included a Kitchen of the Future developed by the Kelvinator division of American Motors, which included an innovative countertop refrigerator. That same year, the new RCA Whirlpool Corporation created a Miracle Kitchen featuring its new products.

Modern American kitchens, a traditional symbol of domestic life, would also become Cold War diplomatic tools, when they began to be featured as centerpieces in cultural exchange exhibits between the United States and the Soviet Union, conducted for decades by the U.S. Information Agency (USIA). The exhibits were technically referred to as "public diplomacy." The first was in July 1959 in Moscow's Sokolniki Park and featured the 1957 Whirlpool Miracle Kitchen, as well as another by General Electric. It was in the Whirlpool kitchen that the famous "kitchen debate" between then-vice-president Richard Nixon and Soviet premier Nikita Khrushchev took place. They enthusiastically and energetically defended their respective country's political systems of capitalism and communism. Khrushchev was contemptuous of American efforts to compete with Soviet accomplishments in space (*Sputnik* was launched in 1958) with kitchens and television sets, which he called "gadgets." But after 32 years of 19 such exhibits viewed by 20 million Soviet citizens, in 1991 the Soviet Union would fail because, among many other things, it could never provide its people with affordable kitchen appliances. Finally, the people would rebel.

I was fortunate to participate in the last of these exhibits in Vladivostok in February 1991, 11 months before the Soviet Union collapsed, as a design expert in an exhibit, Design USA, a showcase of American lifestyles with no need for any verbal or printed political propaganda. Sponsored by the USIA, it attracted nearly 10,000 Soviet citizens per day for one month. With one of the many talented college-age American translators on stage, I described a modern American kitchen to successive waves of visitors, while other design experts in computers, transportation equipment, furniture, or architecture did the same in other exhibits. The kitchen included a range, microwave, washing machine, sink, and various small and major appliances including a large double-door refrigerator/freezer. The kitchen, and especially the refrigerator, was the star of the show. About every 20 minutes, as a new group of Russians assembled in front of the stage, I would announce: "Eta Holodeelnik!" ("This is the refrigerator!"), and would dramatically open the doors while the audience (mostly women) hushed and "oooooed" in awe. This was in February, mind you, when the outside temperature was –30 degrees. Some older women actually had tears in their eyes at that moment. These women were regarding the refrigerator with the same wonder and desire as American women did in the 1930s.

When someone always asked if all Americans had such a large refrigerator, I replied that no, most had smaller models, because there were many choices of sizes in the market, but this size was not uncommon.

Their next question was usually, "well, how do people know what size to buy?" (In the Soviet Union, you were only allowed to live in apartments of a size based on how many were in your family.) I would explain that Americans could choose any size they liked, needed, or could afford, a decidedly foreign concept to most Russians. If I delayed getting around to the refrigerator while explaining other appliances or kitchen features, someone in the audience would always anxiously and helpfully remind me: "Holodeelnik!" "Holodeelnik!" All of this made sense only when I visited with a number of Russian families. Most working class people had no refrigerator at all. In the winter, their freezer was a plastic bag of food hanging out their apartment window. Middle class professionals had the only size refrigerator available on the market, no larger than the 1927 Monitor Top, perhaps three or four cubic feet, which was unremarkable in style or performance. The Russian people were starved for the manufactured fruits of Western mass production, low costs, and high standard of living.

But, I digress. Back to postwar America. Since 1949, when GE experimented with heat pumps, a number of manufacturers developed air conditioners or heat pumps that both heated and cooled. Servel and Fedders had both had introduced one in 1953.[10] In about 1962, General Electric would introduce one called the Zoneline. These units were technically called "packaged terminal air conditioners" (PTACs), "automatic reverse-cycle room air conditioners," or simply "heat pumps." They were originally developed and marketed by companies in the southern United States, where winters were relatively mild. However, many heat pump units were sent up north to be installed in colder climates and unsuitable applications, and the result was a high failure rate that caused a bad reputation. This damaged the market, which would not recover until the 1970s. A more successful solution to year-round heating and cooling was the "rooftop" combination gas heating and electric cooling unit, introduced by the industry in the 2- to 5-ton system range for commercial structures, which enjoyed dramatic growth over the next fifteen years.

By 1955, one in every twenty-two American homes had air conditioning, but in the South, that number was one in ten, twice as many. The colors of domestic air conditioners were uniformly in fairly neutral wood tones or grays, to blend in with the typical room interiors with wood trim. By far, the most popular consumer color, during the 1950s, as researched by the *H&G* Color Program, was gray, and in the 1960s, was beige, often called "almond." But major kitchen appliances followed a more adventurous track, because many housewives preferred a brighter environment for their kitchen. In the 1960s, the *House & Garden* Color Program indicated a decline in popularity of the pastel colors of the 1950s. The next wave of major kitchen appliances, over the 1960s, featured, in sequence, a series of browns, greens, and golds in a two-tone color scheme. Many clothes washers, dryers, stoves and refrigerators were first painted in brown, green or gold, but then, a darker shade of the same color was airbrushed on all edges of the rectangular form, creating a sort of antique, aged appearance. The first of this series in 1964 was a brown color, usually called "coppertone." After a few years, in 1966, the next two-tone scheme was a green, invariably called "avocado," and a few more years after that, in 1968, the most popular was a gold color, often called "harvest gold," or "goldtone." These were all inspired by sequential popular colors in the *House & Garden* charts, described earlier.

In 1958 Frigidaire built its 50-millionth product, its first frost-proof refrigerator/freezer with Frigi-Foam insulation. A major change in the way refrigerators were insulated began in 1960 when Kelvinator produced the first all-foamed-in-place refrigerator. General Electric followed in 1962 with a refrigerator with foamed-in-place urethane. This meant that instead of cutting and trimming pieces of plastic insulation material and assembling them between the metal walls of refrigerators during assembly, which was very labor intensive, the insulation could be injected with a spray gun under pressure, which filled the cavity completely, inexpensively, and quickly. This is just a single example of the many new manufacturing processes that permitted lower costs for consumers. Also in 1962, Henry Uihlein founded the U-line Corporation as an outgrowth of the Ben Hur Manufacturing Company. The company was the first to develop and patent an automatic, stand-alone, under-counter residential icemaker.

In 1963, Westinghouse offered an air conditioner with a "quick mount" installation. Earlier air conditioners weighed over 150 pounds and were quite large and difficult to handle. They had to be installed and mounted by servicemen. Now, with lighter, smaller units, consumers could easily install their own. In 1965, Frigidaire offered its first automatic icemaker refrigerator, which delivered the ice cubes not to a storage container inside the refrigerator, but to an ice saver on the door, for easier access from the outside without opening the door at all. It also included a water dispenser with the same convenience of access. This feature was a huge success in the market place, although initially quite expensive. Innovative decorative options on refrigerators continued. In the 1960s, Nash-Kelvinator debuted its "picture frame" doors on some refrigerator models, allowing owners to match their own kitchen décor by applying decorative fabric or plastic.

Toward the end of the 1960s, there was consolidation of refrigerator manufacturers. American Motors Corporation (AMC) in 1968 sold its Kelvinator Appliance Division operations to White Consolidated Industries (WCI), a sewing machine company founded in 1858 by Thomas H. White. WCI's current sales in 1968 were $830 million. WCI also acquired Frigidaire, Gibson, Tappan, and Franklin Appliance Division of Studebaker, which had lost money in 1964 and 1965, and had turned a pale profit in 1966. Gibson had manufactured refrigerators since 1932, and claimed to have innovated the ubiquitous refrigerator light that automatically goes on when the door opens, and the upright freezer. In 1956, it had been acquired by the Hupp Corporation, which had merged with WCI in 1967.[11] Tappan was a stove manufacturer. Also in 1968, Fedders bought Norge from Borg-Warner for $45 million. Norge sales in 1967 were $114 million. Fedders had been making room air conditioners since 1947. Air conditioning went into space in 1969, when Neil Armstrong and Buzz Aldrin walked on the moon in air-conditioned space suits that protected them from the searing heat of the sun. More than half (54 percent) of new automobiles were equipped with air conditioning in 1969.

Philco-Ford refrigerator/freezer with double vertical doors, 1969, by Alphonse M. Marra and R.E. Munz, vice president (courtesy the Industrial Designers Society of America, Design in America, *1969).*

In 1961, the Ford Motor Company had acquired the Philco Corporation, becoming the Philco-Ford Corporation, which in 1969 introduced a side-by-side, two-door refrigerator/freezer with doors that could easily accept decorative panels to go with any kitchen décor. Designed by Philco-Ford industrial design staff member Alphonse M. Marra and vice president of design R.E. Munz, handles opened respective freezer or refrigerator doors from the center outward, a solution that addressed the right-or-left opening problem. That same year, Westinghouse debuted a refrigerator/freezer designed by Westinghouse industrial design staff members R.W. Kennedy and W.H. Appel under design manager C.F. Graser. It had double French doors with full-length handles for the refrigerator, both opening from the center thus avoiding right-or-left opening problems, and a large pull-out freezer drawer below the refrigerator. Its exterior was customizable with surface treatments such as faux wood paneling, as shown. Also in 1969, General Electric introduced its version of a 23.6-cubic-foot, side-by-side refrigerator/freezer with a unique service bar for ice and chilled water dispenser placed on the outside of one door. Such special convenience features were very popular in families with young children and teens. In 1970, Westinghouse would follow the trend by introducing its refrigerator with a Chill Compartment.

Westinghouse refrigerator/freezer with double French doors (courtesy the Industrial Designers Society of America,* Design in America, *1969).

"Chill compartments" might also be what one could call the place where some humans end up when they die. In the 1960s, the idea to cryopreserve humans came to Robert Ettinger while reading a sci-fi story, and as a result he founded the Cryonics Institute in Clinton Township, Michigan. Inside the institute today, more than 100 people float inside giant bottles filled with nitrogen at temperatures colder than minus 130° Celsius (about −240°F), in the hope that one day in the future, some doctor or new technology will revive them. When Robert died in 2011 at age 92, his son, David Ettinger, explained that his father was

General Electric side-by-side refrigerator/freezer with exterior service for ice and water, 1969 (courtesy the Museum of Innovation and Science, formerly the Schenectady Museum, Schenectady, New York).

temporarily in a "cooling box" preparatory to joining his clients, including Robert's mother, Rhea, his first wife, Elaine, and his second wife, Mae. Robert soon joined them as patient 106. David told ABC news that his father's intention was that he and his family and friends get a chance to live longer and to take advantage of the promise of future technology. David and his wife both believe in his father's ideas, and intend to do the same. "Cryonics is sort of an ambulance to the future," David said.

For approximately $30,000, anyone can be cryogenically frozen. The word "cryogenics" stems from the Greek and means "the production of freezing cold," but today the term is used as a synonym for the low-temperature state. Currently there are over 200 people in a frozen state at cryonics centers around the United States, and some 2,000 people have signed up for the process. Ted Williams, the famous baseball player, had his head frozen in the Alcor Life Extension Foundation in Arizona. Cryogenics is not just for people. Many more things are being frozen, from embryos and umbilical cords to stem cells and semen. The flash freezing of food and biotechnology products, such as vaccines, requires nitrogen in blast freezing or immersion freezing systems.[12] Dermatologists still use liquid nitrogen (at minus 196° centigrade), and cryotherapy equipment (spray cans) to freeze and remove warts, freckles, moles, lesions, age spots, or small skin cancers (the cans sell for $600 to $800 each). And, of course, NASA uses cryogenic fuels such as liquid hydrogen to get space rockets into orbit, so why not use cryogenics to send the deceased into the future? As we will see later, doctors are today bringing people back to life two hours after death, with body cooling.

In 1969, the term "Sun Belt" first began to be used. It refers to the area of the United States that extends from Florida to California and includes industries such as oil, military, and aerospace as well as many retirement communities. Growth in this region had been occurring since World War II, because many military manufacturing jobs had moved from the Northeast to the South and West. But after the Civil Rights Movement, millions of African Americans left the South to find work in the industrial North. Later in the 1960s, as agricultural jobs continued to grow, Mexicans and other Latin Americans began to move north into the Sun Belt. Then, because of intense foreign competition, many Northern corporations began moving their manufacturing facilities to the South, where labor costs were much cheaper and there were fewer labor unions. This also resulted in the migration of Northern employees to manage and supervise the new Southern facilities, and there was rapid growth of the manufacturing industries in the South. Both industry and workers required air conditioning to temper the disadvantage of tropical heat in the summers.

By 1969, most new homes were being built with central air conditioning, and window air conditioners were increasingly affordable. So population in the Sun Belt grew dramatically in the 1970s, largely because of the invention and implementation of affordable and effective air conditioning. The percentage of U.S. population in the South stood at about 24 percent until 1970, but then rose steadily to over 30 percent by 2000.

Without air conditioning, the growth of major cities such as Phoenix, Atlanta, Dallas, Las Vegas, and Houston could not have occurred. The humid and hot summers would have made living and working difficult to bear. The demand for air conditioning in the Sun Belt was also what drove manufacturers of such equipment.

Initially, people were satisfied with a single window air conditioner in the bedroom or living room, but the demand for air conditioning was so great, particularly in the South, that this was not enough. General Electric debuted its first portable air conditioner in 1971, allowing the unit to be easily transported around the house to different rooms as needed. Compared to the costly prospect of buying a half dozen air conditioners for multiple rooms, the idea of a portable conditioner seemed somewhat attractive. But this was less than ideal, because of the hassle of moving the conditioner around the house all day long, the equivalent of carrying a single space heater around the house during the winter. The obvious solution was central air conditioning equipment in the basement or outside that brought cool air throughout the house, and this became more popular and less expensive as time went by.

For large office buildings or hotels, air handling and conditioning is often managed differently. Some systems use water as part of the cooling process. The two most well known are chilled water systems and "cooling tower" air conditioners. In a chilled-water system, the entire air conditioner is installed on the roof or behind the building. It cools water to between 40°F and 45°F and the chilled water is piped throughout the building and connected to air handlers. The water pipes work like the evaporator coils in a standard air conditioner. If the pipes are well insulated, there is no practical distance limitation to the length of a chilled-water pipe. In most air-conditioning systems, air is used to dissipate heat from the compressor coils. In some large systems, a cooling tower is used instead. The tower creates a stream of cold water that runs through a heat exchanger, cooling the hot condenser coils. The tower blows air through a stream of water, causing some of it to evaporate, and the evaporator cools the water stream.

In 1971,Westinghouse completed an experimental house and kitchen in Coral Gables, Florida called Electra '71, designed by the Westinghouse consumer-products design staff. It promoted new concepts in electronic technology, including a cylindrical refrigerator. This was only the latest cylindrical refrigerator concept: the idea has intrigued designers since the 1930s, probably because it solves the problem of easy access to hard-to-reach items in the back of refrigerators, but no models have ever made it to production. The obvious reason is that a cylinder is an incompatible shape in a kitchen filled with rectangular appliances, cabinets, and counters. This is why Westinghouse displayed it as an independent, freestanding concept.

Westinghouse cylindrical refrigerator concept from Electra 71, 1971.

In 1973, Philco introduced a one-piece plastic liner for refrigerators called Cold Guard. Most refrigerators at the time used more than 1,800 kWh (kilowatt hours) per year. A highlight in refrigerator history occurred in 1973 when William Zimmerman of St. Louis got a patent for the first refrigerator magnet. Refrigerators would never be free of kids' drawings again, but at least they were free of Scotch tape. Frigidaire refrigerator models of 1974 included a feature that made ice, ice water, and juice all available without opening the door, but rather by a through-the-door service feature activated simply with pressure on a lever by a glass or cup. Another feature that year was a charcoal deodorizer to remove the odor that accumulates from food. Again, in the 1970s, the color of major appliances shifted in accordance to popular taste, as measured by the *House and Garden* Color Program. While gold and beige were still fairly popular, the overwhelmingly most popular color was a very, very light shade of gray, almost white, and generally called "off white," or "oyster white." Kitchen appliances, including refrigerators, were returning to their traditional commercial category designation of "white goods." Many were white, while others used simulated exotic-wood appliqués of vinyl with chrome trim on the edges.

Refrigerators had by this time become standard household commodities with only relatively minor feature improvements. But they were the activity center of the kitchen amidst many busy lifestyles: where everyone went for a snack or refreshment; where meals were conceived and begun;

or where the refrigerator became a central billboard for kids' drawings, shopping lists, or holiday displays. It is a wonder that some refrigerator manufacturer did not market a model whose door was an actual cork bulletin board with thumbtacks. Nevertheless, in an age of highly mobile populations, 300-pound refrigerators were definitely non-mobile appliances. They were of no use to travelers on a day trip to Grandma's, or for a family trip to the lake or beach for a picnic or sail, or to the stadium for an afternoon tailgate party. These were jobs for the ubiquitous family cooler, a non-electric, insulated, plastic box that served the same purpose and required the same attention as 19th century iceboxes. We still had iceboxes, but now they were portable. Coolers still needed ice to work, but this was now no problem, since home refrigerators produced ice cubes by the basketful, and if that was not enough, bags of packaged ice cubes could be purchased at the local supermarket as the trip started.

Playmate coolers made by the Igloo Corporation and designed by Marlan Polhemus of Goldsmith Yamasaki Specht, Inc., in 1970.

However, the problem with any ice in a box is that it melts when used to keep things cool, and the melt water needs to be drained away. If not, the melt water can accumulate and damage the foods being cooled. For this reason, most coolers have a removable plug on the bottom to drain away ice melt, but in the excitement of picnics and parties, who remembers to do this? Have you ever frozen your hand as you trolled the bottom of an ice-water-filled cooler, feeling for that last bottle of beer or Coke?

What were needed were ice cubes that did not produce water when they melted. Sound impossible? Well, the elegant solution was the "ice pack" or "gel pack," a plastic sack of ice, or of refrigerant gel or liquid. Sealed ice packs contain a liquid that can be frozen in freezers and then transferred to a portable cooler to keep perishable food cool. Water can absorb a considerable amount of heat due to the high enthalpy (latent energy) of fusion of water. When the sealed-in liquid in an ice pack melts or softens, it can be re-frozen and re-used. Similar packs can also be heated to keep food warm when traveling. The first hot and cold pack was introduced in 1948 with the name Hot-R-Cold-Pak; it could be chilled in a refrigerator or heated with hot water.

Gel packs are made of non-toxic materials, such as hydroxyethyl cellulose or vinyl-coated silica gel, which remains as a slow-flowing gel; therefore, it will not spill easily or cause contamination if the container breaks. "Instant cold packs" use an endothermic reaction to cool down quickly. In 1959, Albert A. Robbins patented a "Chemical Freezing Package" that involved an outer pouch containing two separate compartments for water and ammonium nitrate that would mix and freeze when the user split a perforation between the two, say by hitting or snapping the package, or breaking an ampule (sealed plastic or glass bulb). The temperature of the solution fell quickly to about 35°F and could last 100 hours. The Robbins patent was assigned to Kwik-Kold of America, whose parent company, Cardinal Health, continues to market this, and the more recently invented gel-type cold packs, to this day. The first reusable hot-cold ice pack that could be heated in boiling water or heated in a microwave oven was invented in 1971 and patented in 1973 by Jacob Spencer of Nortech Labs, using a flexible gel. Similar packs became useful not only in preserving food, but also in medical and sports fields.

Another type of insulated cooler appeared in the 1970s, one of the many consumer benefits developed during the space age. It is called a thermoelectric cooler. Thermoelectric cooling uses the Peltier effect to create a heat flux between the junction of two different types of materials. The Peltier effect is named after French physicist Jean Charles Athanase Peltier (1785–1845), who discovered the principle in 1834. When direct electric current is applied to such a device, which has two sides, it creates a temperature difference, bringing heat from one side to the other, so that one side gets cooler and the other gets hotter. In a thermoelectric cooler, the hot side is outside the cooler, and the cool side is inside the cooler. The advantage of such a cooler is that it can be powered by connecting to a cigarette lighter in a car or to household electric outlets. Some could also be solar powered. A thermoelectric cooler can typically reduce the temperature by up to 36°F below the ambient temperature. No need for ice cubes or problems with melting water. But they are less efficient in operation compared to vapor-compression systems, such as in refrigerators.

In 1975, White Consolidated Industries (WCI) purchased Westinghouse Electric, creating White-Westinghouse Corporation, and in 1977, WCI also acquired Philco International. That same year, Sears Coldspot refrigerators became part of the Sears Kenmore line and continued to evolve. With the market almost fully saturated, mostly replacement refrigerators were being sold, and the sale of these was directly tied to either remodeling or the new construction of homes. Refrigerators, indeed all major appliances, were no longer seen as freestanding works of art but simply as functional commodities to be integrated invisibly into the overall kitchen architecture and décor. The major driver for the replacement of major appliances in the kitchen was the eternal upgrading of kitchens for the resale of homes, since the kitchen was the key to selling real estate. Although the average life span of a refrigerator is about 17 years, any kitchen older than 10 years was, as any realtor will tell you, a sales deterrent to women, who always want the latest appliances and décor. The same was true in the remodeling trade, when women felt their own kitchens were behind the times and were compelled to upgrade their kitchens. Many Americans at this time were moving their older refrigerators to the garage, to store bulk items and packages of leftovers. The garage fridge became the grungy workhorse of home appliances, sharing space beside the lawn mower, the lawn furniture, and the workbench. But the times were changing, and soon, the refrigeration industry would find itself at the center of global environmental concerns. "Global cooling" would meet what environmentalists would label "global warming."

Chapter 9

Regulations and Climate Change

It began with the energy crisis of the mid–1970s, caused by a shortage of oil caused by an embargo by Middle East oil producers, which triggered a movement toward more efficiency in automobiles and electrical products, particularly refrigerators and air conditioners, which were energy hogs. As a matter of fact, from 1970 to 1975, national annual shipments of air conditioners fell from nearly six million units to three million. Sears introduced its Coldspot Power Miser, touting 40 percent less electricity usage. In the 1970s, Samsung, founded in 1948 and headquartered in Seoul, South Korea, began to make refrigerators. By 1975, Samsung introduced five types of refrigerators, and in 1976, released its energy efficient High Cold refrigerator.

Until the 1970s, chlorofluorocarbons (CFCs), the type of refrigerant used since the 1930s in refrigerators, freezers, and air conditioners, was considered to be non-toxic, non-flammable, and non-reactive with other chemical compounds. Actually, there were dozens of different CFCs with individual chemical specifications varying in the number of chlorine, fluorine, and carbon molecules, some referred to as Freon or Halon (both trade names), some by a range of chemical codes designated as numbers/letters or R-numbers, each having different boiling points ranging from −40°C to +91°C. But in 1973, chlorine was suspected of being a catalytic agent in ozone destruction, a process that removed the odd-numbered oxygen species (atomic oxygen O and ozone O_3) while leaving chlorine unaffected. Conclusive evidence would not be found until 1984, when it would be announced that polar depletion of ozone over Antarctica had been discovered. Ozone protects humans from solar radiation that can cause skin cancer and cataracts. In 1978, the United States banned the use of CFCs in aerosol cans, a major source.

In 1979, Fedders downsized by selling its Norge Division to Magic Chef, a stove manufacturer since 1929, and General Motors sold Frigidaire to White Consolidated Industries (WCI). In 1982, Maytag purchased Jenn-Air, a stove manufacturer that soon would be a refrigerator brand name. In 1984, General Electric came up with an electronic refrigerator that beeped if the door was accidentally left ajar. It was the dawn of the computer age in consumer products. That same year, Apple debuted its Macintosh IIc, and Admiral released an automatic ice cream maker. To use it, one just had to remove the icemaker and slide in the ice cream maker. In 1985, Kenmore released a popular refrigerator with a black lacquer finish. *Time* magazine called the trend "Darth Vaderism," but the popularity of black would increase. In 1986 Maytag absorbed Magic Chef; AB Electrolux of Sweden purchased White Consolidated Industries, including Frigidaire; and KitchenAid was acquired by Whirlpool, which began manufacturing refrigerators under the KitchenAid name.

But refrigerators were now the target of environmental politics. In 1986, Dupont urged worldwide banning of CFCs, and introduced its new HCFCs (hydrochlorofluorocarbons), which still deplete ozone but much less so than CFCs. The ozone friendly refrigerant HFC-R-134a (tetrafluoroethane), synthesized by Albert Henne, co-inventor of the CFC refrigerants in 1936, was designated to replace CFC-containing refrigerants (R-12) in new refrigerators, because of R-134a's lack of chlorine. R-134a also replaced the refrigeration blowing agent, R-11, used to blow polyurethane foam insulation into refrigerators. In 1987, the National Appliance Energy Conservation Act by Congress mandated minimum energy-efficiency requirements for refrigerators and freezers, as well as for room and central air conditioners. Disposal of refrigerators became an increasingly environmental concern, because Freon leakage can damage the ozone layer. Disposal of refrigerators became regulated, often requiring the removal of doors. Children had been asphyxiated while hiding inside discarded refrigerators. Since 1956, under U.S. federal law, refrigerator doors were no longer were permitted to lock from the inside. More modern units use a magnetic door-gasket that holds the door shut but can be pushed open from the inside.

Also in 1987, a United Nations international treaty, the "Montreal Protocol on Substances that Deplete the Ozone Layer," intended to begin phasing out CFC refrigerants, was opened for signatures, and was entered into force on January 1, 1989, ratified by 197 nations, including all members of the United Nations. The U.N. protocol called for a complete phase-out of HCFCs by 2030, and it was believed that if this international agreement were adhered to, the ozone layer would recover by 2050. However, in 2006, NASA reported the largest hole in the Antarctic ozone on record,

10.6 million square miles. Later, in 2007, when carbon dioxide became the primary target of environmentalists, 200 countries would agree to eliminate HCFCs earlier, by 2020. Later, ozone-friendly R-134a also became a target for future phase-out because it still contributed to global warming with carbon dioxide, although much less so than CFCs. By 2012, it was being reported that the ozone hole had decreased to the smallest size since 2002, although it is constantly changing. Environmentalists regarded the reduction as positive evidence that man can, in fact, modify global climate through the regulation of offending chemicals. Currently in Europe, R-134a is becoming uncommon, and the main new refrigerant being used is R-600a (isobutene).

Meanwhile on the design side of the industry, so-called post-modern design made an appearance in the mid–1980s, as some designers rebelled against the uniform appearance of many commodity consumer products such as kitchen appliances and computers. Architects such as Robert Venturi and Michael Graves had initiated the movement in the 1970s, characterizing the absence of ornamentation in the purely functional glass-box International Style architecture established by the Bauhaus in the 1920s and the Museum of Modern Art in the 1950s as "bland, sterile, characterless, or worse, hostile to its environment and humanity." Architects began to violate these traditional design rules by incorporating visual references to historical design elements, and featuring ambiguity and playfulness in visual elements of their work.

It was Italian designers who first applied post-modern philosophy to product design in 1981, when members of the Milanese Studio, Alchimia, founded Memphis, a group of avant-garde designers. Memphis created the design media event of the year with garish and irreverent furniture designs in bold colors and laminates that emphasized surface decoration more than function. The designs were colorful, cheerful, playful, non-architectural, non-functional, preposterous, and faddish, and all were presented to the beat of rock music in the manner of a fashion show. Post-modernism was picked up in U.S. industrial design educational institutions and re-labeled "product semantics," which avoided the geometric "black box" character of many modern designs, and incorporated humanistic or animalistic metaphors, visual puns, sensuous forms, and whimsical references. Major appliance manufacturers in Europe began to produce radios that looked like motorcycles and a microwave oven that looked like a nuclear reactor. Despite skepticism from mainstream designers, post-modernism inspired designs that were more playful, colorful, and dramatic than "black boxes," the derogative term post-modernists applied to modernist designs.

Refrigerators, since the 1950s, had been pretty similar in outward appearance for reasons mentioned earlier, that is, having to do with fitting into modern, coordinated kitchens as a modular component. They had also become "white boxes," the esthetic equivalent of electronic "black boxes." By the 1980s, the *House and Garden* Color Program had pretty much ended with the retirement of Faber Birren as its consultant, and most refrigerators were again in traditional white, or sometimes black. Refrigerators thus became a target for Italian designers. In 1987, an Italian designer, Roberto Pezzetta, designed a post-modern refrigerator for Zanussi, a major Italian major appliance manufacturer, which challenged not only the standard form of a refrigerator, but its typical white color, as well. The Zanussi model was not commercially successful, since its dramatic shape and color did not fit in with traditional kitchens.

Zanussi refrigerator by Robert Pezzetta, 1987.

Also in 1987, the National Appliance Energy Conservation Act mandated minimum energy-efficiency requirements for refrigerators, freezers, and room and central air conditioners. The legislation was prompted by the oil embargos implemented in the 1970s by Middle East oil providers, creating an unstable market and increasing energy costs. The United States initiated programs to make the country less energy dependent. In 1992, the U.S. Energy Policy Act mandated similar energy standards for commercial buildings. Such energy reduction policies are still in effect today, and many industries still focus on energy reduction in their manufacturing, and in the products they manufacture.

At least one engineer was challenging refrigerators in the 1980s—not for their energy use, their engineering, or even their appearance, but for their operating instructions, which, of course, are part of the design intended to instruct the user, essentially a human factors issue. Donald A. Norman, an accomplished MIT engineer, in his 1988 book *The Design of Everyday Things*, described his ordinary, two-compartment refrigerator, complaining that he couldn't set the temperature properly. His refrigerator had two controls, one for the freezer compartment and another for the food compartment, which suggested that there were two thermostats, one in each compartment. The instructions were to adjust both settings for normal, colder or coldest, and then allow 24 hours for the temperatures to stabilize. In reality, there was only a single thermostat in an unknown location, which adjusted the cooling unit to produce more or less cold air in general, not for separate compartments. The cold air was then sent to both the freezer and the food compartment through a single tube that split into a "Y," of which one leg went to the freezer and the other to the food compartment. In the "Y" was a pivoting valve, which was the other adjustment made by the user when setting the controls. The valve simply directed the amount of the single stream of cold air that went to the freezer or the food compartment, more or less to one or the other. Some of today's temperature controls on similar units are the same, and equally frustrating for users to adjust. The problem seems to remain.

Norman's point was that user controls can often confuse more than they can help, because the user cannot see or understand what the hidden mechanisms are actually doing. The user is misled by the controls, which incorrectly indicate an oversimplified version of what is actually taking place. As mechanical devices become more complex, designers and engineers struggle to devise controls that the user will perceive as simple and easy. This is a noble objective but can easily lead to user confusion when the results do not match what the user expects (today's computerized controls of appliances and automobiles take this problem to a new level of frustration).

A similar confusion arises because of the hidden and extremely complex mechanisms in the earth's biosphere that determine global cooling or warming. Scientists try to simplify these so that the public can better understand how to control the atmosphere, optimistically assuming that this is technologically possible. Man's confidence in technology seems to be infinite. By 1988, some scientists were becoming concerned about the rise of carbon dioxide in the atmosphere. This was their reasoning: they knew that to stay at a constant temperature, the earth had to radiate as much energy at it received from the sun. They calculated that a planet at our distance from the sun, emitting the same total amount of energy it receives, would have a temperature well below freezing. But it is not, because infrared radiation, beaming up from the earth, is intercepted by "greenhouse" gas molecules in the lower atmosphere that keep the lower atmosphere and the surface warm. The nitrogen and oxygen molecules that make up some of the greenhouse gases do not intercept infrared radiation. The most important greenhouse gases are natural water vapor (about 95 percent) and other gases, mostly man-made: carbon dioxide (less than 1⁄10 of 1 percent), methane (about 6⁄100 of 1 percent) nitrous oxide (about 5⁄100 of 1 percent), and others (about 5⁄100 of 1 percent). Scientists observed that carbon dioxide was rising rapidly, and the only reasonable explanation for this, they concluded, was that it was due to human activities. They calculated that if no steps were taken to curb such emissions, by the end of the 21st century the temperature of the earth would rise by 1°C (1.8°F). These conclusions raised serious questions about potential future global disasters such as the ice caps melting and rising sea levels around the world that could destroy major population centers.

The threat of what would soon become popularly called "global warming" aroused the environmentalists. Environmentalism, which looked to and lobbied governments to address environmental issues with legislation, had been growing since the 1950s. In 1955, the Air Pollution Control Act provided research and technical assistance relating to the control of air pollution, which was considered to be a danger to public health. In 1963, Rachel Carson's book *Silent Spring*, which is credited by many with launching the environmental movement, focused public awareness on the environmental effects of pesticides, including DDT, as well as the health hazards of many pesticides to humans. Then, in 1966, Ralph Nader published *Unsafe at Any Speed*, which cited the danger of automobiles to the environment by their emissions and their lack of safety due to poor engineering design and neglect by auto manufacturers.

Due to the growing public awareness fostered by Carson's and Nader's books, the Air Pollution Act was amended in 1967 to become the Air Quality Act, which gave states the right to enforce federal automobile emissions standards. In 1970, as an annual Earth Day was initiated, amendments to the Air Quality Act completely re-wrote the act, creating the U.S. Environmental Protection Agency, which allowed citizens to sue polluters or government agencies for failure to abide by the act, and required that by 1975, the entire United States would attain clean-air status. The Clean Water Act was passed in 1972, and the same year the famous photo of the earth called "Blue Marble in Space" was taken from space; it not only popularized the environmental movement, but also symbolized the idealism of many environmentalists for whom eliminating pesticides and reducing auto emissions were not grand or all-inclusive enough goals. They were inspired to assume the responsibility for the largest and most inclusive environmental task possible to imagine, that of "saving the world." Today, that slogan convinces many to feel they are contributing to a just cause. A common current supermarket sign urging customers to stop using plastic bags states: "Save the world, one bag at a time."

The Energy Policy and Conservation Act was passed in 1975, and environmental organizations were founded throughout the world. By the time the ozone hole in the Antarctic was confirmed in 1986, the environmental movement had evolved an effective global template to address such potential disasters with international conferences that influenced governments, and the 1988 finding by scientists predicting global warming was a ready-made global cause for environmentalists. The United Nations (UN) led the way. In 1988, it established the Intergovernmental Panel on Climate Change (IPCC), charged with producing regular reports on the subject. Four years later, in 1992, the UN created a treaty, the United Nations Framework Convention on Climate Change (UNFCCC or FCCC). Its ambitious objective was to "stabilize 'greenhouse gas' concentrations in the atmosphere at a level that would prevent dangerous anthropogenic [human] interference with the climate system." To do so, it would negotiate specific international treaties (called protocols) that could set limits on greenhouse gasses. Actually, the only greenhouse gas targeted was carbon dioxide, the gas presumably caused primarily by humans. Thus began the international pressure to take global action against carbon dioxide and the industries that caused it.

The United States implemented Energy Star, a government-backed program established in 1992 under the Clean Air Act, "helping businesses and individuals protect the environment through superior energy efficiency." Current U.S. appliance models that are Energy Star qualified use 50 percent less energy than average models made in 1974. There are subdivisions of Energy Star criteria. Tier 1 refrigerators are 20 percent to 25 percent more efficient than the federal minimum standards set by the National Appliance Energy Conservation Act of 1987. Tier 2 are those that are 25 percent to 30 percent more efficient, and Tier 3 is the highest qualification, for those at least 30 percent higher than federal standards.

About 80 percent of Energy Star–qualified refrigerators are Tier 1, 13 percent as Tier 2, and 5 percent at Tier 3. Also in 1992, the U.S. Energy Policy Act mandated minimum energy-efficiency standards for commercial buildings, using research and standards developed by the American Society of Heating, Refrigerating, and Air Conditioning Engineers (ASHRAE). Many of the refrigerators made in the 1930s and 1940s were far more efficient than most made later. This is partly true because of the later addition of new features such as automatic defrost, and ice dispensers and water chillers that became popular in the 1970s. Because of new energy-efficiency standards, refrigerators are now much more efficient than those made in the 1930s; they consume the same amount of energy while being three times as large.

Returning to the mundane world of the refrigeration business, we find that Robur Group, an Italian company founded in 1956, acquired the gas air-conditioning division of Servel, assuring the distribution of its products in America. In 1992, Danby Products began U.S. operations in Findlay, Ohio. Danby was founded in 1947 in Montreal, Canada, manufacturing small electrical appliances. During the late 1970s, it had begun importing a line of compact chest freezers from Italy. In 2011–2012, Danby created an office in China and acquired other companies to begin supplying Coldtech commercial grade refrigerators and freezers to the restaurant industry. Coldtech had previously only sold directly to the end user. In 1992, General Electric (GE) introduced the first complete line of stylish, integrated appliances specifically designed to rejuvenate kitchens. In 1994, GE debuted the first 30-cubic-foot, freestanding, side-by-side refrigerator, the world's largest, and in 1996, GE followed this with the largest-capacity "built-in" style refrigerator. In 1995, Sub-Zero challenged the idea of the refrigerator as one centralized appliance with the debut of its 700 series: multiple pullout cooler drawers that customers could place throughout a room. Philco introduced its AquaPure ice and water filtration system in side-by-side refrigerator/freezers in 1998. Frigidaire followed suit with its PureSource ice and water filtration in side-by-sides. In 1999, both Kelvinator and Frigidaire claimed to produce 70 percent of all U.S. freezers.

Around the turn of the century, watching celebrity chefs on television, who were invariably using professional major appliances of stainless steel, consumers wanted the same look in their own kitchens. GE complied with this new demand by introducing a line of stainless steel domestic appliances in 1999. In 2000, GE introduced its Profile Arctica refrigerator with CustomCool. The system contained special bins, controlled by dampers, a fan, a temperature thermistor, and a heater, which could be used to quickly chill or thaw items, or hold a certain temperature. It automatically reported time remaining to complete the desired function, but did not affect temperature in the rest of the refrigerator. Typical refrigerators are now divided into four temperature zones to store different types of food:

Freezer:	−18°C (0°F)
Meat zone:	0°C (32°F)
Cooling zone:	5°C (41°F)
Crisper:	10°C (50°F)

That same year, Amana introduced a curved front on its refrigerators to add a more organic look to rectangular, architectural boxes. Such curved features appear on a number of major appliances, the result of post-modern design, mentioned earlier. 2001 triggered a flurry of major refrigerator introductions, apparently to celebrate the new century. Kelvinator debuted its Home Smart Refrigerators, a totally redesigned and engineered line; White Consolidated Industries introduced its Next Generation line under the Frigidaire brand name; and AB Electrolux debuted its Advance Tech line. In 2002, LG Electronics, founded in Seoul, South Korea, in 1958 (the name LG is the initials of the two companies that merged in 1958, Lak-Hui and GoldStar), launched the world's first "Internet" refrigerator, also called a "smart" refrigerator, which was programmed to sense what kind of products were being stored inside it and keep track of the stock through barcode or RFID (Radio-frequency identification) scanning. Goldstar began exporting electronic products to the United States in 1962. In 2002, Amana was sold to Maytag.

Refrigerators and all other emitters of carbon dioxide, however, were again at the forefront of world affairs. Since the 1992 U.N. treaty to halt human generation of carbon dioxide, hundreds of scientists with government grants had been busy amassing statistics, measurements, and predictions of global warming to support the UN's claim. In 1997, the UN Framework Convention on Climate Change (FCCC) negotiated the Kyoto Protocol, an amendment that would commit countries to reduce their emissions of carbon dioxide and five other greenhouse gasses (water vapor, methane, nitrous oxide, ozone and CFCs). However, since carbon dioxide was the major gas controllable by humans, it was the primary target of the UN.

Further, the Kyoto Protocol "recognized that developed countries are principally responsible for the high levels of emissions in the atmosphere as a result of 150 years of industrial activity." Therefore, developed nations were committed to "binding targets" of reduction in emissions, while so-called developing countries did not have binding targets; in fact, their emissions are allowed to grow "in accordance with their development needs." There were also complicated formulas enabling trading of emission quotas among countries, and acquiring credits by financing emission reductions in developing countries. These activities would, of course, involve financial exchange. In fact, the Protocol requires that developed countries have to pay billions of dollars and supply technology to other developing countries for climate-related studies and projects. Most developing countries, the vast majority of UN member states, were pleased with the treaty, as they had no obligations, and pleased with the fact that developed nations were the ones paying for their industrial sins of the past.

Not unexpectedly, U.S. approval of the Kyoto Protocol became a political hot potato. It had to be approved by both the Senate and the executive branch. Before the Protocol was agreed upon, the Senate in 1997 passed a resolution unanimously (95–0) disapproving of any international agreement that "(1) did not require developing countries to make emission reductions, and (2) would seriously harm the economy of the United States." Therefore, although Bill Clinton's administration signed the treaty, it was only symbolic and was never submitted to the Senate for ratification.

When George W. Bush was elected U.S. president in 2000, he said that he took climate change "very seriously," but that he opposed the Kyoto treaty, because "it exempts 80% of the world, including major population centers such as China and India, from compliance, and would cause serious harm to the U.S. economy." These same concerns would later become fiscal realities for other developing countries that had approved the treaty. In 2011, Canada, Japan, and Russia stated that they would not take on any further Kyoto targets. Its costs of implementation were prohibitive. Indeed, to meet its theoretical targets, the United States would have had to make specific commitments to dramatically reduce its usage of coal and oil, basic energy sources for many industries, and replace them with more eco-friendly sources such as wind, atomic, and solar energy. Such changes would require the costly conversion of power plants and traditionally powered vehicles, the massive funding of research and development of alternative sources, and the loss of thousands of jobs in labor-intensive industries such as coal and oil.

The Democrats, whose constituency included most environmentalists, and who embraced the idea of new energy sources and climate control in their progressive agenda, were furious with President Bush's decision to not approve the Kyoto agreement. Al Gore, the Democratic vice president under Bill Clinton and Democratic candidate for president who had lost to George Bush in 2000, was particularly enraged, because he was a staunch advocate of the UN's push for control of the earth's climate to combat global warming. He was further frustrated by the 2004 presidential election, in which George W. Bush was re-elected over the Democratic candidate, John Kerry. In 2005, and without U.S. participation, the Kyoto Protocol came into effect following its ratification by Russia.

Meanwhile, the refrigeration industry was interested in cooling, not warming. In 2002, Whirlpool unveiled its Polara, an oven that doubles as a refrigerator. Before leaving for work in the morning, you just throw a par-cooked meal into the unit, which is chilled in default mode. When activated by a timer or cell phone, the refrigerator shuts off and turns the oven on to finish the meal. In 2003, the United States produced over 11.5 million refrigerators and China produced nearly 30 million. South Korea was third, with over 7 million. Among these was LG Electronics, which introduced the first TV as a built-in feature in its refrigerator, and in 2007, upgraded that feature to high definition TV, along with weather center, photo slideshow, digital recipe card, and calendar functions. In 2004, Kelvinator introduced a line of 60-cm- (2-foot)-deep built-in refrigerators and the Big Family Pair, a refrigerator and upright freezer paired together for 946 liters (33.5 cubic feet) of storage capacity. Frigidaire debuted 60-cm- (2-foot)-deep refrigerators with "Precision Air Filtration and Easy Care Real Stainless Steel." In 2005, Frigidaire introduced a new color, Silver Mist, which had the look of stainless steel, and debuted the Prestige Pair, a refrigerator and upright freezer, a kitchen centerpiece with 946 liters (33.5 cubic feet) of fresh and frozen food storage. Sub-Zero, which specializes in built-in refrigerators, debuted its 600 series, a line of refrigerators with glass doors. They saved energy and provided a feeling of openness in a room, but had to be tidied up whenever guests arrived. Sub-Zero refrigerators were usually wider than most refrigerators, to make up for the less depth (24 inches) required matching normal cabinet depth. They were also heavier, sometimes twice as heavy (800 pounds) as a regular refrigerator (250–300 pounds).

General Electric in 2005 introduced its Monogram Wine Vault, which held 1,100 bottles of wine, the first with electronic Cellar Management System, an inventory system with a 15" touch screen, bar code screener, label printer and software to manage, store, and protect liquid assets. Is it a refrigerator or a computer? It's both. By 2005, after the more than 90 years that refrigerators had been on the market, 99.5 percent of all U.S. households owned at least one.

In 2006, Frigidaire introduced the first frost-free chest freezer, and Electrolux unleashed the Cyber Fridge, a wireless–Internet-surfing, e-mail checking, MP3-playing, touch-screen fridge. The screen was located above the door so one person could scour the Internet while another foraged for food. Whirlpool acquired Maytag Corporation, which included the Jenn-Air and Amana brands. Foreign competition in refrigeration surged in 2006, when Daikin Industries, Ltd., a Japanese multinational air conditioning manufacturing company headquartered in Osaka with operations in Japan, China, Australia, India, Southeast Asia, Europe, and North America, acquired OYL Industries, making it the second largest HVAC manufacturer in the world after the Carrier Corporation. OYL had been incorporated in Shah Alam in 1974, assembling gas cookers and ovens, and later became a member of the Hong Leong Group, Malaysia. Daikin had been founded in 1924 to produce aircraft radiator tubes and produced its first refrigerator in 1934. After World War II, in 1951, it produced packaged air-conditioner systems, and in 1994 established Daikin America, Inc. manufacturing in the United States. Daikin co-developed a R-410A refrigerant with Carrier, and was an innovator in the Split System Air Conditioning market and the inventor of Variable Refrigerant Flow (VRF) AC systems. VRF is a technology that uses refrigerant as the heating and cooling medium, and allows one outdoor condensing unit to be connected to multiple indoor fan-coil units (FCU), each individually controllable by the user, while modulating the amount of refrigerant being sent to each evaporator. The system works by allowing individual units to heat or cool as required locally, and can save up to 55 percent in energy.

In 2007, vintage-style refrigerators (with modern engineering) returned with the renewed interest in mid-century "modern" design. Notable companies were Gorenje, Big Chill, and Smeg. Designers showed off a modular refrigerator at the annual Electrolux Design Lab. Split into transparent, stackable compartments, it used eco-friendly technology called magnetic refrigeration. It was more practical than another concept fridge that looked like a tree, also displayed at the Design Lab.

Saving energy was at the top of Barack Obama's Democratic political agenda when he was elected president of the United States in 2008, implementing environmental controls despite U.S. rejection of the global Kyoto Protocols, and despite economic concerns regarding the recession and high unemployment. He and his administration gave billions of dollars to the wind, solar, and geothermal industries through both direct subsidies and in the form of tax credits, but failures such as Solyndra have cost taxpayers more than $500 million. The administration cut greenhouse gasses from cars and light trucks, demanding that by 2025 the U.S. auto fleet must average 54.5 miles per gallon.

Through the Environmental Protection Administration (EPA), coal and oil-fired power plants were required to control emissions of mercury and other poisons for the first time. The rule affected about 40 percent of the power plants that lack modern pollution controls. So-called cap-and-trade legislation was intended to reduce carbon dioxide emissions by capping an industry's allowable emissions but allowing them to trade (buy) credits from other industries that were well below their own caps. A House of Representatives bill was passed to this effect in 2009, but died in the Senate, controlled by Democrats.

Saving energy continued to be a high priority in the refrigeration industry. In 2010 General Electric introduced its Energy Star–qualified, 29.1-cubic-foot side-by-side refrigerator, with 18 cubic feet of storage for fruits, vegetables, and dairy products, and an 11.1-cubic-foot freezer. Most refrigerators today use less than 500 kWh (kilowatt hours) per year, compared to the 1,800 kWh that was common in 1974 refrigerators. Sears in 2010 introduced a refrigerator that used the same amount of energy as a 75-watt light bulb. That same year, Jenn-Air introduced an integrated Built-in Refrigerator Collection. By 2011 Whirlpool was marketing not only Whirlpool appliances but a number of other brand names including Maytag, KitchenAid, Jenn-Air, Amana, Brastemp, Consul, and Bauknecht. Whirlpool's annual revenues were $18 billion. By 2012, General Electric marketed an Ecomagination Suite of kitchen appliances, including a refrigerator, to reduce energy consumption in an average kitchen by up to 20 percent.

In 2012, there was a new world leader in HVAC systems. In 2008, Daikin had purchased a 75 percent share of All World Machinery Supply to develop hybrid hydraulic systems, a product intended to cut energy consumption. In 2012, Daikin acquired Goodman Global for $3.7 billion, which was expected to expand Daikin's presence in the United States and to make Daikin the world's largest maker of HVAC systems. Other refrigerator brands on the U.S. market alone today, but not mentioned earlier, include Keystone, Equator, and Kenmore brands, sold through Kmart; Avanti compact refrigerators; Midea compacts; Marshall fridge; EdgeStar; Koldfront; and Polar (commercial).

Refrigerators today come in a range of colors. Brand name refrigerator colors have generally become limited to highly popular standards that have been fairly consistent for several decades. The most popular finishes are probably white, black, and authentic stainless steel, all essentially neutral in character. Black became popular in the 1980s and stainless steel in the 1990s. In 2012, General Electric introduced a warm gray color called slate, which looks somewhat like stainless steel, and which became very popular. For the more adventuresome, there are some lines of various wood grain decorative finishes that blend into a wood paneled kitchen to appear as built-in appliances, and for others, who like the look of the '50s and '60s, there are some manufacturers who specialize in retro-looking designs in a rainbow of bright colors such as red, yellow, green, or blue. Some people remove the door and have it painted in a custom color at an auto body shop. Of course there are also collectors who shop on-line for actual antique models that still function, to have a completely authentic retro look.

Making refrigerators in a variety of colors and finishes is relatively easy for manufacturers, but coping with federal and international regulations is another story. Such regulations started with the phasing out of CFCs in 1986 to reduce ozone depletion and the adoptions of HCFCs, which reduce ozone, resulted in the United Nation's Montreal Protocol of 1989, which will ban the current HCFCs completely by 2030. In 1987 the U.S. Congress enacted the National Appliance Energy Conservation Act to conserve energy in all major appliances. Many current models reduce energy by 50 percent. By 1988, scientists were predicting global warming by using computer simulations, and by 1992, the United Nations Intergovernmental Panel on Climate Change (IPCC) created a treaty to set limits on "greenhouse gases." Carbon dioxide was the target of the UN's 1997 Kyoto Protocol to reduce greenhouse gases. The threat of global warming has been diminished somewhat, since the ICPP chairman admitted in 2014 that world temperature records have remained relatively flat for the last 17 years.

Global warming may seem to have little to do with the refrigeration industry, but as in many industries, government regulations to address perceived threats that are 50 to 100 years in the future add layer after layer of costs to research, development, and manufacturing, which is often passed on to consumers of products like refrigerators. And often, such perceived long-term threats take decades to be verified as to their reality or not, long after the government administrators who initiated them, and those who oppose them, are gone. There are several basic connections of climate change to refrigeration:

First, if global warming actually turns out to be occurring, whether by natural or human means, the refrigeration industry will be booming like never before and it will become an essential mechanism for human survival. So if you believe in global warming, you should probably invest in refrigeration. Second, believe it or not, some people are actually thinking of cooling the entire planet by sucking heat-trapping CO_2 from the air, or reflecting sunlight back into space. It's called geo-engineering, and is considered Plan B, since Plan A, slashing carbon dioxide emissions from fossil fuels, is moving so slowly. The Intergovernmental Panel on Climate Change (IPCC), in its draft document for a new global climate pact in 2015, says that unless emissions are cut much faster than currently projected, measures to scrub CO_2 from the air will have to be deployed to avoid potentially dangerous levels of warming. Plan B would remove CO_2 from the air and store it underground, or solar radiation could include everything from covering open surfaces with reflective materials or placing sun-mirrors in orbit around the earth. The problem with these exotic technologies is that they don't exist yet, or are in an experimental stage. No one knows whether they will be successful.

Now, let's review some final reflections on our close friend and companion, our local, family kitchen refrigerator. Today, we regard our refrigerator as just another convenient household appliance that we scarcely think about until it needs repair or replacement. We no longer worry about its function, its appearance, or its household safety, but only about its energy consumption and its effect on the environment. However, like many of our modern conveniences, it is the result of generations of scientists, inventors, engineers, promoters, and designers who, bit by bit, refined and improved it to modern perfection. The history of the refrigerator is typical of the transformation of our society from an agrarian age to an industrial age, and to an electronic age. We are the beneficiaries of technology.

Refrigerators, the most ubiquitous and familiar refrigeration device to many people, may have reached their apogee of development and refinement. For over 200 years, since Thomas Moore's portable icebox "refrigerator" of 1800, they have evolved into mechanical marvels that are the center of our kitchens and casual lifestyles. It is hard to imagine a television beer commercial without some pointed reference to refrigerators, ice, or snow. Dagwood Bumstead's life seems to revolve around his family refrigerator. But, as technological development continues, refrigerators will inevitably improve. Perhaps there will be computer refinements such as inventorying the contents and automatically reordering food as needed, or requiring users to identify themselves to open the door (when on certain diets). History tells us that we cannot discount the possibility that current refrigeration systems will eventually be replaced by some magical new, yet-unknown technology. Current needs in the market are based on two factors: energy conservation and environmental impact.

For example, a new type of air conditioning for cars, expected to be introduced in 2015, is called TIFFE (Thermal systems Integration For Fuel Economy), which is said to reduce gas consumption by 15 percent. Another example: the Stratos Company LLC is working on a novel R718 air conditioner that is expected to be very inexpensive. It purports to use a refrigerant known as R718, which is more commonly known as water vapor. It is claimed to be more efficient than current refrigerants, such as R134A, but takes a special type of compressor to make it work. The compressors used in Europe, where energy costs are relatively high, are made of titanium turbines, because R718 compressors have to spin very fast to get the right pressures. Developers are experimenting with relatively cheap carbon-fiber plastic instead of titanium.

A new project at the National Physical Laboratory (NPL) and Imperial College are working to make the inefficient (maximum 10 percent efficiency) method of refrigeration and air conditioning obsolete, using the electro-caloric effect to develop new methods of cooling. The electro-caloric effect is a phenomenon in which a material changes temperature under an applied electric field. It is said to require less energy than the compression process to create the same level of cooling. It would also reduce size and weight, and would eliminate the chemicals that harm the environment. The leading scientist at NPL is confident that by 2016, they can develop the first electro-caloric refrigerator ever to operate at close to room temperature.

There is also something called magnetic refrigeration, an effect first observed by German physicist Emil Warburg in 1881. Its fundamental principle was suggested by Dutch-American physicist Peter Debye (1884–1966) in 1926, and by American chemist William Giauque (1895–1982) in 1927. The first working models were constructed in 1933. The magneto-caloric effect is a phenomenon in which a change in temperature of a suitable material is caused by exposure to a changing magnetic field. Such suitable materials include the chemical element gadolinium and some of its alloys. In August 2007, the Rise National Laboratory at the Technical University of Denmark claimed to have reached a milestone in this technology, and hoped to introduce the first commercial applications in 2010. Cooltech has been trying to commercialize this technology since 2011, and at the end of 2013, a production line was completed with a capacity of 10,000 units per year. Other scientists predicted in 2005 that within 10–15 years, the technology could be available in home refrigerators and air conditioners. General Electric is working on similar systems, current versions of which are 20 percent more efficient than today's systems and can reduce temperatures by 80 degrees. GE predicts that they will hit the consumer market by 2020.[1]

Other potential new systems could be much closer. For example, solar energy is currently being used to feed power into the electrical grid and thus power conventional refrigerators. There is no reason that the heat generated by solar collectors could not right now directly power absorption refrigeration and air conditioners, and in fact such examples have been in use in the United States and Australia for decades.[2]

Refrigeration as it is today can perform more spectacular tasks than the typical preservation of food and making of ice in the lowly and underappreciated household refrigerator. A current proposal under consideration illustrates how it could protect the environment and human life from deadly contamination. You may recall the catastrophic failure at the Japanese Fukushima Daiichi Power Plant in March 2011, resulting in a meltdown of the plant's six nuclear reactors. It was caused when the plant was hit by the tsunami caused by an earthquake. No one died because of the small amounts of radiation, but the evacuation of 300,000 people in the local area resulted in 1,600 deaths from evacuation conditions such as temporary housing and hospital closures. The most dangerous remaining threat was the extensive leaking of contaminated ground water from the reactor and storage tanks through the soil into the sea. On August 19, 2013, a huge leak of contaminated water—about 80,000 gallons—was discovered.

The proposed solution to the problem was the construction of an underground ice wall around the reactor to contain the seepage. Coolant pumped through pipes at −20°F to −40°F reaching 100 feet deep into the soil would freeze any seepage. The ice wall would keep clean groundwater from coming into the plant, and water from radioactive particles from getting into the ocean. Similar applications could solve similar biohazards throughout the manufacturing industries.[3]

Refrigeration plays an important role in saving lives that few people are aware of. Back in 1972, Raymond Damadian developed the first magnetic resonance imaging machine (MRI). In 1980, a Scottish team used an MRI machine to produce the technology's first diagnosis of cancer. The breakthrough would not have happened without helium; the powerful MRI magnets cannot function without it. But helium for this purpose must be in liquid form, and to get it to be a liquid, you have to get it really cold. Liquid helium is the coldest substance on earth: −452.2°F (this is actually near absolute zero, which is –459.67°F). When everything else is frozen solid, helium is still a liquid. That specific chemical property makes it an invaluable resource. It is essential to bring the winding wire on a magnet below what is called its critical temperature, the point at which the wire becomes not just a conductor of current, but a superconductor.

Dutch scientist Heike Kamerlingh Onnes first discovered superconductivity in 1911 by liquefying helium. He won the Nobel Prize in physics in 1913 for his work, which led directly to the research that gave us MRIs. The stronger the currents conducted through the wire by superconductivity, the stronger the magnetic field. The field interacts with the atoms that make up every substance—such as the human body—to give a detailed image of the structure within. Doctors can now see tumors growing inside the human body, and scientists can examine the substances that treat those tumors. Mechanical refrigeration is the only way to turn helium into a liquid, and therefore is essential to enable the medical field to perform the miracle technology of MRI,[4] but this can only be achieved using the Hampson-Linde cycle for refrigeration, a technique using temperature changes induced by compression and decompression of a gas.

Cryonics, the study of the production and behavior of materials at very low temperatures, is also used to freeze and preserve male sperm and female eggs, so that these can be preserved for as long as 21 years, thus enabling men with vasectomies to still have children at a later time, and women with cancer to preserve their eggs before chemotherapy or radiotherapy so that they can have children later in life. This emerging medical technology is properly known as cryonics, and often erroneously referred to as cryogenics in the popular press. Supercooling is most commonly used for rocket fuels, for example, liquid hydrogen, to launch space vehicles. During World War II, it was discovered that metals frozen to low temperatures showed more resistance to wear. In 1966, Ed Busch founded the first cryogenic processing industry, called CryoTech, which experimented in increasing the life of metal tools to anywhere from 200 percent to 400 percent of the original life expectancy through cryogenic tempering instead of heat treating.

Scientists debate at what point on the temperature scale refrigeration ends and cryogenics begins, but the commonly accepted point is at –180°C (about –292°F) This is a logical dividing line, since the normal boiling points of the so-called permanent gases (such as helium, hydrogen, neon, nitrogen, oxygen, and normal air) lie below –180°C while the Freon refrigerants, hydrogen sulfide, and other common refrigerants have boiling points above –180°C.

Although amazing in many ways, mechanical refrigeration has been only one of the many technological amenities we enjoy today that were the result of scientists, inventors, developers, engineers, and designers who struggled to refine their creative visions into reality. It is now time to go to your refrigerator to refill your glass of ice tea with "cubes" and to see what frozen delicacy you are having for dinner tonight. After dinner, check out the next chapter to see how natural snow and ice is still surviving today. Stay cool!

Chapter 10

Epilogue: Snow and Ice Redux

Most of this book has been about the development of mechanical refrigeration, which made possible the manufacture of artificial ice and subsequently, the manufacture of modern refrigerators, freezers, and air conditioners. In conclusion, it seems only fitting that we return to the subject of natural snow and ice, which inspired man's search for these mechanical replicas of nature.

Ancient Romans only valued snow as a way to cool their wine. Today, natural snow serves an important scientific purpose as a frozen history of deep time. By examining ice cores taken from ice sheets formed by layers of snow over thousands of years in polar regions of the globe, scientists can reconstruct a climatic record through isotopic analysis. Depths of cores can be as much as 100 meters. Inclusions in the snow and ice each year remain there, such as wind-blown dust, ash, bubbles of atmospheric gas, and radioactive substances; and temperatures over time can be reconstructed by analyzing the annual layers of snow. Cores have been obtained representing climate changes over the past 420,000 years. They have recently provided reliable historical temperature and carbon dioxide data to scientists, so that they can predict global warming (or cooling, as the case may be) more accurately.

As a matter of fact, snow and ice are essential factors in calculating global climate change. Together, they comprise the cryosphere, the portions of the earth where water is in a solid form, including sea ice, lake and river ice, snow cover, glaciers, ice caps, and ice sheets. The greater part, 68.7 percent, of the world's freshwater is in the form of ice and snow, and 1.9 percent of the world's salt water. Most of the earth's ice mass—90 percent—is in Antarctica, and 10 percent in the Greenland ice cap. Most of the annual snow cover is in the Northern Hemisphere, and variability is dominated by the seasonal cycle. During January in the Northern Hemisphere, snow and ice cover 23 percent of the hemispheric area, or 18 million square miles, but in August, a mere 3.5 million square miles. Northern American winter snow cover has shown an increasing trend over much of this century, but satellite data show there has been little annual variability over the 1972–1996 period.

Sea ice, or frozen salt water, varies also by season but differs between hemispheres. For example, the overall trend from 1978 to 1995 in Arctic sea ice decreased 2.7 percent per decade, while during the same period, Antarctic sea ice increased by 1.3 percent per decade. Relationships between global climate and changes in ice extant are complex. Most of Antarctica never experiences surface melting, and may in fact be taking water out of the oceans. It is a possibility that global warming could result in losses to the Greenland ice sheet being offset by gains to the Antarctic ice sheet. But those concerned about global warming fear the collapse of a floating ice shelf, such as the Ross ice shelf, which, they predict, would raise the world sea level 15 feet over a few hundred years.

They also point to the recession of glaciers worldwide. While glacier variations are likely to have minimal impact on global climate, their recession may have contributed one-third to one-half of the observed 20th century rise in sea level, mostly in Western North America, where runoff from glacierized basins is used for irrigation and hydropower. A clear understanding of the mechanisms at work is crucial to interpreting the global change signals that are contained in the glacier mass balance records. As glaciers increase in size and advance to lower elevations, the higher temperatures at those lower levels cause more melting. Often, the disappearance of the lower portion reduces melting, thereby increasing mass balance (the difference between winter snow accumulation and the amount removed by summer melting) and potentially establishing equilibrium over a few decades.

With such complexities of global ice variations, future estimates of ice extant carry an uncertainty of 20 percent. It is easy to see how scientists can vary enormously in their interpretation and computer model calculations, depending on their pessimism or optimism, their grant providers, or on their political leaning.

Ice can preserve the past in many ways. One would think that by the 1990s, there would have been no more icemen, but suddenly, forty years after most genuine icemen (the guys that delivered ice to iceboxes in homes) disappeared, one reappeared from the past. The oldest iceman in the world (well, not really a traditional iceman), but nevertheless called "Ötzi the Iceman," is a well-preserved natural mummy of a man who lived around 3,300 BC, and who was discovered in 1991, frozen in ice in the

Ötzal Alps on the border between Austria and Italy. He is on display, along with his clothing and possessions, in a cool exhibit, refrigerated of course, in the South Tyrol Museum of Archaeology in Bolzano, South Tyrol, Italy. For nine Euros, you can see him Tuesdays through Sundays, or daily in July, August, and December.

Natural ice has preserved not only 5,300-year-old humans, but also extinct animals, such as a remarkably intact, 40,000-year-old baby mammoth, Baby Lyuba, that was discovered in May 2007 on the banks of the Siberian Yuribel River. The 39,000-year-old remains of a baby female woolly mammoth from the last ice age was discovered in 2012 on the frozen New Siberian Islands off Russia, with much of her fur and trunk intact. She was named Yuka, and was about two and one half years old when she died. Scientists managed to obtain samples of mammoth blood at the Museum of Mammoths at Yakutsk, Siberia, which they hope can be used to clone woolly mammoths. Yuka went on display at an exhibition hall in Yokohama, near Tokyo, Japan, in 2013. Also in May of 2012 a fully grown female mammoth from 50 to 60 years of age, and weighing about a ton, was found frozen in ice on the Lyakhovsky Islands, the southernmost group of the New Siberian Islands in the Arctic Seas of northeastern Russia. During life, she would have weighed about three tons. This was the third such adult mammoth carcass that has been found in the history of paleontology.

Although no longer widely used for domestic refrigeration, natural ice and snow are a fascination to many people around the world who use it enthusiastically for recreational activity. If you watched the 2014 Winter Olympics in Sochi, Russia, you saw dozens of such perennial sports: ice hockey, figure skating, alpine skiing, ski jumping, snowboarding, curling, cross country, bobsleds, short track—the list goes on. And this is not to mention individual, mostly non-competitive sports such as snowmobiling, dog-sledding, ice sailing, ice fishing, and mountain climbing.

Many of these sports can be traced to prehistoric times. A wooden ski dating from earlier than 5,000 BC was found about 1,200 miles northeast of Moscow at Lake Sindor, and other ancient archeological evidence of skis have been found in Sweden, Norway, and even in Roman-era Italy, where Mount Etna on Sicily is covered in snow during the winter. Modern skiing evolved in Scandinavia for purely utilitarian travel on snow, but in the mid–1800s became a popular pastime and sport. The first skiing competition in Norway was held in 1843. In 1861, the first cross-country skiing club was formed in Norway, and that same year the first alpine skiing club was formed in Australia. The first public ski-jumping competition was held in Trysil, Norway, in 1862, and the oldest ski club in North America was the Nansen Ski Club, founded in 1872 by Norwegian immigrants in Berlin, New Hampshire. When the Winter Olympics started in 1924, the five winter sports were bobsleigh, curling, ice hockey, Nordic skiing (which included military patrol, cross-country, Nordic combined, and ski jumping), and skating (including figure skating and speed skating). Mogul skiing and freestyle skiing were added to the 1992 Winter Olympics.

Sled dogs probably evolved in Siberia when humans migrated north of the Arctic Circle with dogs and began using them to pull sleds about 3,000 years ago, to hunt, fish, and travel. Historical references indicate dog harnesses were used by American Indian cultures before European contact. The use of dogs as draft animals was widespread in America. Sled dogs were used to deliver the mail in Alaska during the late 1800s and early 1900s, each team hauling 500 and 700 pounds of mail. Mail delivery ended in 1963, when the last carrier, Chester Noongwook of Savoonga, retired. In 1925, there was a diphtheria outbreak in Nome, Alaska. There was not enough serum in Nome to treat people infected by the disease. There was serum in Nenana, 700 miles away, but it was only accessible by dogsled. So a dog-sled relay was set up in the villages between the two towns, and 20 teams worked together to deliver the serum to Nome in six days. The Iditarod Trail was established on the path between these two towns, Iditarod being the largest village on the trail.

Primitive ice skates appear in the archeological record from about 3000 BC in the form of animal-bone skates found on the banks of Lake Moss, Switzerland. Ice-skating is mentioned in a book from the 1100s in England, describing skates made of bones, which were used with long sticks to propel them faster. The Dutch in the 13th or 14th century devised steel skates with sharpened edges, when canals connecting towns were frozen for months and skating was the fastest means of travel. The Dutch developed a classic skate design during that era that remains unaltered to this day. In 1891, the first European figure-skating championships were held in Hamburg, Germany. Figure skating was the first winter sport included in the Summer Olympics, in 1908.

Ice hockey probably evolved from field hockey, which had been played for centuries in northern Europe. In the 1100s, children with bone skates pushed with sticks to increase their speed. Some attacked each other competitively with the sticks until one fell down, often breaking arms or legs. All that was needed was a puck, and perhaps a loose stone became one. The rules of modern ice hockey were devised in Canada in 1875, when the first game was played with these rules in Montreal. The first artificial ice rink was built in 1876 at the Glaciarium in London. For 66 years, the only way to resurface ice rinks was to scrape, wash, and squeegee the ice by hand, then add a thin layer of water to fill blade cuts and to refreeze a fresh surface. Then in 1949, Frank J. Zamboni (1901–1988), originally in the refrigeration business, invented the Model A Zamboni Ice-Resurfacer, a machine that accomplished this elaborate process much faster and more easily. Original Zamboni machines were gasoline engine–powered, but these became too dangerous to use indoors due to the carbon monoxide produced. In 1959, Zamboni developed an electrically powered machine that was used for the Squaw Valley Olympic Games in California in 1960.

Curling was developed in medieval Scotland, when stones on ice were used in 1541 at Paisley Abbey, Renfrewshire. The word "curling" first appeared in print in Perth, Scotland, in 1620. Unlike those of today, the original playing stones were simply flat-bottomed river stones of inconsistent size, shape, and smoothness. The first curling club in the United States was established in 1830; the first world championship for curling was in 1959 in Falkirk and Edinburgh, Scotland; and it has been an Olympic event only since 1998.

Ice yachting was in vogue in 1790, on the Hudson River, with headquarters in Poughkeepsie, New York. Ice yachts were simple, square boxes on three runners, with the rear runner operated like a tiller for steering. By 1879, the model for all Hudson River ice yachts was the *Icicle*, a 69-foot structure with over a thousand square feet of canvas sail. Today's ice yachts consist of a 40-foot backbone, with a runner plank at right angles to it supporting the two forward runners, and a 3-foot rear rudder runner at the rear of the backbone. Sails are generally 400 square feet. Depending on the wind, ice yachts can achieve speeds from 70 to 90 miles per hour.

Bob sledding, or bobsleighing, was invented by the Swiss in the late 1860s. Teams made timed runs down narrow, twisting, banked iced tracks in a gravity-powered sled. The sport, however, did not begin until the late 19th century, when the Swiss attached two skeleton sleds together and added a steering mechanism. A chassis was added to give protection to wealthy tourists. The world's first bobsleigh club was founded in St. Moritz, Switzerland, in 1897. By the 1950s, the critical importance of the fast start had been recognized, and athletes with explosive strength were drawn to the sport. In 1952, a critical rule change limiting the total weight of crew and sled ended the era of the super heavyweight bobsledder and rebalanced the sport as an athletic contest.

Snowboarding is the most recent snow sport. As early as 1910, people would tie wooden planks to their feet to steer themselves downhill on snow. Modern snowboarding began in 1965, when Sherman Poppen, an engineer in Michigan, invented a toy for his daughter by fastening two skis together and attaching a rope to one end to steer as she stood on the board and glided downhill. It was called the "snurfer," combining "snow" with "surfer." Poppen licensed the idea to a manufacturer that sold about a million "snurfers" in the next decade. Various entrepreneurs perfected the design during the 1970s, and the first competition offering prize money was the National Snurfing Championship at Muskegon State Park in Michigan in 1979. The first world championship half-pipe competition was held at Soda Springs, California, in 1983. Early snowboards were banned from ski slopes, but by 1990 most ski areas had separate slopes for snowboarders. In 2012, the International Paralympic Committee announced that adaptive snowboarding (dubbed "para-snowboarding") would debut as a men and women's medal event in the 2014 Paralympic Winter Games, which took place in Sochi, Russia.

Natural snow for centuries has been used by Inuit American Indians as construction material for igloos (Inuit word: *iglu*) in Canada's Central Arctic and Greenland's Thule area. A skilled Inuit can construct an igloo, a dome-shaped, livable, temporary structure within an hour, and indeed, it can become a comfortable home in the coldest of weather, because snow, with air pockets trapped in it, is a natural insulator. On the outside, temperatures may be as low as –49°F, but on the inside, temperatures may range from 19°F to 61°F when warmed by body heat alone. When a stove, candle, oil lamp or other heating device is used, the igloo will hold that heat and increase the interior temperature accordingly. Snow used to build an igloo must have enough structural strength to be cut and stacked appropriately. The best is snow that has been blown by wind, which can compact and interlock ice crystals. Independent blocks can be cut with a long, sharp tool, polished to fit each other closely, and laid in a spiral, circular pattern, gradually smaller, to form a dome. In some cases a single block of clear ice is inserted to allow light inside. In others, a short tunnel is constructed at the entrance to reduce heat loss when the door, or hide, is opened. In emergency situations, when travelers have been trapped by sudden snowstorms, many lives have been saved by makeshift snow caves or crude snow tunnels, which provided shelter and afforded protection from the cold.

A recent example occurred in March 2013, when 17-year-old Nicholas Joy of Medford, Massachusetts, tried to find a shortcut through the woods while skiing and became lost west of Sugarloaf Mountain in Maine. Nicholas's father, Adam Joy, knew his son was missing when he didn't meet him in the ski resort parking lot as planned. The Sugarloaf Ski Patrol, the Maine Forest Service, the U.S Border Patrol and other area rescue squads and volunteers searched for Nicholas for two days and two nights, though weather conditions were bad enough at night that the search had to be suspended. Nichols, recalling survival-oriented reality–TV shows he had watched, made himself a shelter out of a mound of snow and tree branches. He also tried to start a fire by rubbing two sticks together. Nicholas had no food or cell phone with him. A volunteer fireman who was not part of the official search parties found the teenager the morning of the third day. Nicholas was tired, hungry, and cold, but unharmed. "He did the right thing in building a snow cave," Mr. Joy told the Associated Press. Obviously, he was still alive to talk about it, so he made some good decisions.

Natural ice is used not only as an architectural medium but also as an artistic medium. In 1989, the world's first ice hotel was built in the Swedish village of Jukkasjävi. That year, Japanese ice artists visited the area and created an exhibition of ice art. In spring of 1990, French artist Jannot Derid held an exhibition in a cylinder-shaped igloo nearby. One night there were no rooms available in the village of about 548 residents, so some of the visitors asked permission to spend the night in the exhibition hall. They slept in sleeping bags on top of reindeer skins, and were the first guests of the "hotel." The entire hotel is made out of snow and ice blocks taken from the Tome River.

Each spring in March, the Icehotel harvests ice from the frozen Tome River and stores it a nearby production hall holding more than 100,000 tons of ice and 30,000 tons of snow. The ice is used for creating an ice bar that serves drinks in ice glasses, while the snow is used for building a strong structure for the building. Each year for the past 24 years, the Icehotel has accepted 200 applications from artists to design and build an "art suite" and to host events and product launches all over the world.

Other ice hotels were built. In Japan, the Alpha Resort Tomamu in Shimukappu village on Hokkaido Island builds an ice hotel each year, which includes a 15-meter dome where everything from the bed to the table and ceiling is one big ice sculpture. The hotel interiors are typically at 28°F to 30°F, and guests dine on ice tables and sit on ice chairs covered with sheepskin. In Canada the Hôtel de Glace (Ice Hotel) opened in 2001. It was originally located on the shores of Lac-Saint-Joseph, Quebec, 31 minutes north of Quebec City, but later moved closer to the city, 10 minutes from Old Town Quebec. In 2006, Kirkenes Snow Hotel opened in the easternmost town in Norway, close to the Russian border. It has 20 rooms, decorated by ice artists from Finland and Japan. Also in 2006, a 14-room ice hotel and an adjacent ice church were built in the Fagaras Mountains in Romania, where ice sculptors became imitators of the modernist Romanian sculptor, Constantin Brancusi (1876–1957).

Ice sculpture has been an international art form for many years. Favorite subjects are swans and doves for wedding receptions. Cruise ship buffets are famous for their use of ice sculptures, as are many prominent hotels and restaurants. There are schools and textbooks for ice carving, a skill often taught in culinary schools to enhance the presentation of food. Traditional ice-carving tools are razor-sharp chisels and hand saws, but more recent methods use power tools, including chain saws. The latest techniques include computer-driven cutting machines and standard molds. Clear ice is favored by sculptors, and after carving, the surface can be heated with a propane torch to create a fully transparent look.

A number of international ice and snow sculpting events are held throughout the world, mostly in countries with cold winters. Quebec City in Canada, which initiated ice sculpting in 1880, holds an ice sculpture festival each year during the Quebec City Winter Carnival, featuring an ice castle. Twenty teams are chosen to participate. Each January, the Fairmont Chateau Lake Louise, at Lake Louise in Banff National Park, Alberta, Canada, hosts a three-day event called Ice Magic, sanctioned by the National Ice Carving Association. Twelve teams of three carvers are given fifteen blocks of ice, weighing 300 pounds each, and must transform them into ice sculptures within three days. In northeast China, Heilongjiang Province hosts an International Ice and Snow Sculpture Festival in Harbin, the largest in the world. It began in 1963 but was interrupted by the Cultural Revolution, and resumed in 1985. Ice sculpture has been popular in Japan since the 1930s. The Japanese city of Sapporo on the Island of Hokkaido conducts an annual winter carnival that features creative ice sculptures the size of multiple-story buildings. Since 1989, Alaska has hosted the annual World Ice Art Championships, and one hundred sculptors come from around the world to compete. Teams compete in the Single Block Classic (7,800 pounds) competition or in the Multi-Block Classic (ten blocks of 4,400 pounds each). More than 45,000 spectators attend. The annual London Ice Sculpting Festival, the largest in England, began in 2009.

In addition to these highly publicized and professional sculpture competitions, when it snows each winter throughout the northern states we can find literally thousands of amateur snow sculptures in front yards from Massachusetts to Montana. Wet snow seems to beg to be transformed into a three-dimensional representation of winter more creative than the classic snowman. In an average winter of half a dozen snows, some families will build six different snow sculptures, light them with spotlights, and amaze their neighborhood as effectively as with traditional strings of Christmas lights. The process can become a family or neighborhood team effort. Depending on the age, experience, and creativity of the artists, results can range from your standard, three-ball snowmen with nose carrot and muffler, to the most exotic animal sculptures imaginable.

Of course, this is not to say that natural snow and ice are always welcome or pleasurable. Winter weather can be downright frustrating, with roads closed, air traffic halted, and elevated heating bills. Snow and ice can be dangerous, with falling tree limbs, loss of electricity, and car accidents. Nothing has spurred the population move to southern states over the last 50 years more than a growing distaste for northern winters. Retirees relocating to Florida love nothing more than watching northern

Snow sculpture, February 2014 (courtesy Robert E. Schott, Cranford, New Jersey).

winters play out on television news, but they are thankful for their air conditioners in July and August.

Most recreational activities using natural snow and ice occur outdoors in far northern or winter climates, with natural snow and ice, but you don't have to wait for winter, cold weather, or nature. Today man can make his own ice and snow to replicate nature. There are thousands of indoor, artificially produced ice rinks for both competitive sports and recreational use. There are also literally dozens of indoor, climate-controlled, year-round ski resorts operating around the world, with artificially produced snow. One of the most unusual is located in, of all places, Dubai, an emirate in the United Arab Emirates Federation, on the southeast coast of the Persian Gulf in the Middle East, where summer highs average about 106°F with overnight lows of about 86°F and winters have an average high of 75°F, with overnight lows of 57°F. If you thought this was not a good place for a ski resort, you'd be wrong.

In this desert region is an indoor ski area as part of the Mall of the Emirates, one of the largest shopping malls in the world. The indoor resort, opened in 2005, features an 85-meter-high, 22,500-square-meter, indoor mountain with five slopes of varying steepness and difficulty, including a 400-meter-run, the world's longest indoor run. A quad chairlift and tow-lift carry skiers and snowboarders up the mountain. Adjoining the slopes is a 3,000-square-meter play area with sled and toboggan runs, an icy body-slide, climbing towers, giant snowballs and an ice cave. It is also home to a number of penguins who come out to play several times a day. Winter clothing and snowboard equipment are included in the price of admission. An extremely efficient insulation system helps the facility maintain a temperature of 30°F during the day and 21°F at night, when snow cannons manufacture the snow.

Art Hunt, Dave Richey, and Wayne Pierce invented the snow cannon, a device that manufactures snow, in 1950, and went on to patent it in 1952. The first resort in the world to use artificial snow was Grossinger's Catskill Resort Hotel in New York State. Snow cannons began to be used on a commercial scale in the 1970s. Snowmaking begins with a water supply such as a river or reservoir. The water is pumped up a pipeline on a mountain, and distributed through an intricate series of valves and pipes to any trails that require snowmaking. Many resorts add a nucleating (crystallization) agent to insure that as much water as possible freezes and turns to snow. The next step is to use an air plant, which is usually a building containing electric or diesel industrial air compressors the size of a van or truck, although some have portable, trailer-mounted compressors. The air is generally cooled and excess moisture removed before entering the system, but in some systems the water is cooled, as well. Although there are many different types of snowmaking guns, they all share the basic principle of combining air and water to form snow. For most guns, you can change the type or "quality" of snow by regulating the amount of water added to the mixture.[1]

Of course, you can make snow yourself. All you need is cold weather and a pressure washer. Use a fine-mist nozzle with cold water (the colder the better), and point it away from you at a 45-degree angle. Or use a garden hose with a snow nozzle. It makes less snow than a pressure washer, but it's still fun. Want snow in warm weather? Make fake snow by adding water to sodium polyacrylate, a gel that is inside disposable diapers, as crystals to keep soil moist in a garden center, or from a chemical supply house. Add more water (it will not dissolve) and get wetter snow.

Despite the fact that virtually all homes now have ice cube machines in their refrigerators, the manufactured, bagged ice industry is still a $2.5-billion-a-year Goliath. Local mom-and-pop operations have been replaced by mega-distributors such as Reddy Ice, Polar Ice, or Arctic Glacier. But the basic delivery mechanism of ice plant production distributed on a refrigerated truck delivered to a refrigerated on-site cooler has remained unchanged for over 60 years. However, this system has recently been hit by increasing labor costs, insurance, and fuel costs. Therefore, to avoid transportation costs, some ice manufacturers such as Polar Ice have converted to producing and packaging ice at the point of sale with low-maintenance ice machines that can produce 12,000 pounds of ice per day. So we may return to the local manufacture of ice. This is quite a contrast to the 1800s, when natural ice was transported around the world by ship and yet was still highly profitable.

Ice can do much more than treat bruises and swelling from bumps or bites; it can save human lives. Donated human organs must be quickly transplanted into a recipient, usually within 24 hours or less, depending on the organ. During organ-donation surgery, just prior to the removal of a donated organ, an ice-cold preservative solution containing electrolytes and nutrients is flushed into the organ to remove the blood. This begins the process of preserving the organ. Sterile ice is also placed in the body cavities to aid in the cooling. Once the organs are removed from the donor's body, they are packed into sterile containers and further cooled when these containers are surrounded with an icy slush mixture. The goal is to cool but not freeze the organs. Organs other than kidneys are stored using simple hypothermia, that is, kept cold in the preservation solution.

Kidneys are placed on a machine called a pulsatile perfusion device, which continuously pumps preservation solution through them, and can be preserved up to 48 hours.

There's more. For decades, it was common knowledge in treating patients with cardiac arrest that if the heart stopped beating for longer than six to ten minutes, the brain would be dead. Now, a new treatment being used by a growing number of hospitals suggests that a patient can be brought back to life even if the heart is stopped for twenty minutes or longer. It is a significant difference. Many patients who would previously been given up for dead have been revived and discharged with all or nearly all of their cognitive abilities intact.

The treatment is called therapeutic hypothermia, and uses the most simple of technologies: ice. Once the patient's heartbeat is restored, doctors, cardiologists and rescue squads apply ice to moderately lower a patient's body temperature by about six degrees. The patient is then put into a drug-induced coma for 24 hours before being gradually warmed back up to normal temperature. According to the Minneapolis Heart Institute, an early adapter to the procedure, cardiologists have treated 140 patients since 2006 and 52 percent have survived compared to single digits historically, and of those, 75 percent reported a favorable recovery, and many reported a full return to normal activity. Not an insignificant result from using nothing but frozen water!

At the Presbyterian Hospital in Pittsburgh, Pennsylvania, in March 2014, doctors were planning to save the lives of ten patients with knife or gunshot wounds by placing them in suspended animation, buying time to fix their injuries. Neither dead nor alive, the victims were to be cooled down by replacing all their blood with a cold saline solution, which rapidly cools the body and stops almost all cellular activity. "If a patient comes to us two hours after dying, you can't bring them back to life. But if they are dying and you suspend them, you have a chance to bring them back after their structural problems have been fixed," said surgeon Peter Rhee at the University of Arizona in Tucson, who helped develop the technique. "We are suspending life, but we don't like to call it suspended animation because it sounds like science fiction," said Dr. Samual Tisherman, a surgeon at the Pittsburgh hospital, who was leading the trial. "So we called it 'emergency preservation and resuscitation.'"

The procedure differs from therapeutic hypothermia, described above, because when someone reaches the emergency room with a traumatic gunshot or stab wound, slow cooling is not an option. Often their heart has stopped beating due to extreme blood loss, giving doctors only minutes to stop the bleeding and restart the heart. The patient will have lost about 50 percent of their blood. Their chance of survival is less than 7 percent. In the new procedure, cold saline is flushed through the heart up to the brain, the area most vulnerable to low oxygen. To do this, the lower region of the heart must be clamped and a catheter placed into the aorta, to carry the saline. The clamp is removed and the saline pumped around the whole body. It takes about 15 minutes for the patient's temperature to drop to about 50 degrees F. At this point, they are clinically dead. The patient is taken to surgery to fix the body injuries, and because of the slowing of body metabolism, the doctors have about two hours. The saline is then replaced with blood.

Getting this technique into hospitals was not easy. Because the trial would occur during a medical emergency, neither the patient nor their family could give consent. The trial could proceed only because the U.S. Food and Drug Administration considered it to be exempt from informed consent. The technique was to be tested on ten people, and the results compared to another ten who met the criteria but who were not treated this way because the team wasn't on hand. The technique was to be refined, then tested on another ten, until there are enough results to analyze.

So ice and snow today are still a source of human enjoyment, scientific study, artistic creativity, and medical breakthroughs, as well as still filling their earliest human use of providing cooling and comfort. Most importantly, ice and snow were the initial inspiration for humans to make artificial ice, which over many years led to the mechanical refrigeration, freezing, and air conditioning that enhance our present lives.

Chapter Notes

Chapter 1

1. Nagengast, "A History of Comfort Cooling Using Ice," 49.
2. Donaldson et al., *Heat and Cold*, 19.
3. James and Thorpe, *Ancient Inventions*, 323.
4. Donaldson et al., *Heat and Cold*, 18.
5. James and Thorpe, *Ancient Inventions*, 322.
6. Donaldson et al., *Heat and Cold*, 18.
7. James and Thorpe, *Ancient Inventions*, 322.
8. Ibid.
9. Donaldson et al., *Heat and Cold*, 18.
10. James and Thorpe, *Ancient Inventions*, 320.
11. Donaldson et al., *Heat and Cold*, 18.
12. Messadié, *Great Modern Inventions*, 16.
13. James and Thorpe, *Ancient Inventions*, 321–22.
14. Donaldson et al., *Heat and Cold*, 18.
15. Wikipedia, "Ice Cream," http://en.wikipedia.org/wiki/Ice_cream (accessed June 6, 2014).
16. Messadié, *Great Modern Inventions*, 16.
17. Donaldson et al., *Heat and Cold*, 34.
18. Ibid., 45.
19. Ibid., 36–38.
20. Ibid., 45.
21. Ibid., 21.
22. Ibid., 23.
23. William Black as quoted in Stradley, "Ice Cream History."
24. Mrs. Mary Eales, *Mrs. Mary Eales's Receipts*, Recipe # 93 (London: Printed for J. Robson, 1733), http://www.gutenberg.org/files/20735/20735-h/20735-h.htm (accessed May 7, 2014).
25. "Beethoven," from Godey's Lady's Book 41 (August 1850), as quoted in Stradley, "Ice Cream History."
26. Ibid.
27. Ibid.
28. Donaldson et al., *Heat and Cold*, 46.
29. Ibid., 47.
30. Ibid., 45–46.
31. Wikipedia, "Frederic Tudor," http://en.wikipedia.org/wiki/Frederic_Tudor (accessed May 13, 2014).
32. Wikipedia, "Ice Trade," http://en.wikipedia.org/wiki/Ice_trade (accessed May 13, 2014).
33. Ibid.
34. Wikipedia, "Nathaniel Jarvis Wyeth," http://en.wikipedia.org/wiki/Nathaniel_Jarvis_Wyeth (accessed May 13, 2014).
35. Wikipedia, "Ice Trade," http://en.wikipedia.org/wiki/Ice_trade (accessed May 13, 2014).
36. Wikipedia, "Nathaniel Jarvis Wyeth," http://en.wikipedia.org/wiki/Nathaniel_Jarvis_Wyeth (accessed May 13, 2014).
37. Wikipedia, "Wenham Lake," http://en.wikipedia.org/wiki/Wenham_Lake (accessed May 13, 2014).
38. Derry and Williams, *A Short History of Technology*, 698.
39. Wikipedia, "Ice Trade," http://en.wikipedia.org/wiki/Ice_trade (accessed May 13, 2014).
40. Rees, *Refrigeration Nation.*
41. Ibid.
42. Wikipedia, "Frederic Tudor," http://en.wikipedia.org/wiki/Frederic_Tudor (accessed May 13, 2014).
43. Donaldson et al., *Heat and Cold*, 46.

Chapter 2

1. Wikipedia, "Icebox," http://en.wikipedia.org/wiki/Icebox (accessed March 21, 2014).
2. Wikipedia, "Ice Trade," http://en.wikipedia.org/wiki/Ice_trade (accessed May 13, 2014).
3. Donaldson et al., *Heat and Cold*, 51–52.
4. Wikipedia, "Ice Trade," http://en.wikipedia.org/wiki/Ice_trade (accessed May 13, 2014).
5. Beecher, "Care of the Cellar," in *The American Woman's Home; or, Principles of Domestic Science*, chapter 30.
6. Messadié, *Great Modern Inventions*, 17.
7. Donaldson et al., *Heat and Cold*, 43.
8. Macauley, *The Way Things Work*, 160.
9. Donaldson et al., *Heat and Cold*, 43.
10. Wikipedia, "Benjamin Franklin," http://en.wikipedia.org/wiki/Benjamin_Franklin (accessed June 25, 2014).
11. Donaldson et al., *Heat and Cold*, 118.
12. Ibid., 117.
13. Wikipedia, "Jacob Perkins," http://en.wikipedia.org/wiki/Jacob_Perkins (accessed May 15, 2014).
14. Wikipedia, "Nicolas Appert," http://en.wikipedia.org/wiki/Nicolas_Appert (accessed May 13, 2014).
15. Donaldson et al., *Heat and Cold*, 117.
16. Ibid., 262.
17. Ibid.
18. Ibid., 268.
19. Ibid.
20. Wikipedia, "Dry Ice," http://en.wikipedia.org/wiki/Dry_ice (accessed May 13, 2014).
21. John Gorrie as quoted in Burke, *Connections*, 239.
22. Burke, *Connections*, 238–39.
23. Donaldson et al., *Heat and Cold*, 119.
24. Ibid., 120.
25. Ibid.
26. Ibid., 117.
27. Ibid.
28. Ibid., 119.
29. Burke, *Connections*, 240.
30. U.S. Patent 8,080.
31. Donaldson et al., *Heat and Cold*, 124.
32. Ibid.
33. Wikipedia, "Air Conditioning," http://en.wikipedia.org/wiki/Air_conditioning (accessed February 12, 2014).
34. Donaldson et al., *Heat and Cold*, 266.
35. Ibid., 124.
36. Ibid., 126.
37. Ibid.
38. Wikipedia, "Refrigeration," http://en.wikipedia.org/wiki/Refrigeration (accessed May 13, 2014).
39. Donaldson et al., *Heat and Cold*, 126.
40. Wikipedia, "James Harrison," http://en.wikipedia.org/wiki/James_Harrison_(engineer) (accessed May 13, 2014).
41. Donaldson et al., *Heat and Cold*, 126.
42. Wikipedia, "James Harrison," http://en.wikipedia.org/wiki/James_Harrison_(engineer) (accessed April 15, 2014).
43. Ibid.
44. Donaldson et al., *Heat and Cold*, 127.
45. Derry and Williams, *A Short History of Technology*, 698.
46. Donaldson et al., *Heat and Cold*, 144.
47. Ibid., 147.
48. Wikipedia, "Lager," http://en.wikipedia.org/wiki/Lager (accessed May 13, 2014).
49. Burke, *Connections*, 243.
50. Wikipedia, "Lager," http://en.wikipedia.org/wiki/Lager (accessed May 13, 2014).
51. Wikipedia, "Thomas Sutcliffe Mort," http://en.wikipedia.org/wiki/Thomas_Sutcliffe_Mort (accessed May 13, 2014).
52. Donaldson et al., *Heat and Cold*, 239.
53. Wikipedia, "Ferdinand Carré," http://en.wikipedia.org/wiki/Ferdinand_Carré (accessed May 13, 2014).
54. Wikipedia, "Refrigeration," http://en.wikipedia.org/wiki/Refrigeration (accessed May 13, 2014).
55. Donaldson et al., *Heat and Cold*, 134.

56. Ibid.
57. Ibid., 135–36.
58. Texas State Historical Association, "Refrigeration," http://www.tshaonline.org/handbook/online/articles/dqr01 (accessed January 18, 2014).
59. Wikipedia, "Refrigeration," http://en.wikipedia.org/wiki/Refrigeration (accessed May 13, 2014).
60. Donaldson et al., *Heat and Cold*, 130.
61. Texas State Historical Association, "Refrigeration," http://www.tshaonline.org/handbook/online/articles/dqr01 (accessed May 13, 2014).
62. Daniel Somes as quoted in Donaldson et al., *Heat and Cold*, 267.
63. Donaldson et al., *Heat and Cold*, 268.
64. Nagengast, "A History of Comfort Cooling Using Ice," 49.
65. Donaldson et al., *Heat and Cold*, 268.
66. Nagengast, "A History of Comfort Cooling Using Ice," 50.
67. Donaldson et al., *Heat and Cold*, 269.
68. Wikipedia, "Ice Trade," http://en.wikipedia.org/wiki/Ice_trade (accessed May 13, 2014).
69. Ibid.
70. Ibid.
71. Donaldson et al., *Heat and Cold*, 138–39.
72. Wikipedia, "Thaddeus S. C. Lowe," http://en.wikipedia.org/wiki/Thaddeus_S._C._Lowe (accessed May 13, 2014).

Chapter 3

1. Texas State Historical Association, "Refrigeration," http://www.tshaonline.org/handbook/online/articles/dqr01 (accessed May 13, 2014).
2. Burke, *Connections*, 241.
3. Texas State Historical Association, "Refrigeration," http://www.tshaonline.org/handbook/online/articles/dqr01 (accessed May 13, 2014).
4. Wikipedia, "James Harrison (Engineer)," http://en.wikipedia.org/wiki/James_Harrison_(engineer) (accessed May 13, 2014).
5. Burke, *Connections*, 242.
6. Wikipedia, "James Harrison (Engineer)," http://en.wikipedia.org/wiki/James_Harrison_(engineer) (accessed May 13, 2014).
7. Donaldson et al., *Heat and Cold*, 139.
8. Messadié, *Great Modern Inventions*, 17.
9. Burke, *Connections*, 242.
10. Donaldson et al., *Heat and Cold*, 130.
11. Derry and Williams, *A Short History of Technology*, 699.
12. Wikipedia, "Refrigeration," http://en.wikipedia.org/wiki/Refrigeration (accessed March 4, 2014).
13. Wikipedia, "Ice Rink," http://en.wikipedia.org/wiki/Ice_rink (accessed March 4, 2014).
14. Donaldson et al., *Heat and Cold*, 147.
15. Texas State Historical Association, "Refrigeration," http://www.tshaonline.org/handbook/online/articles/dqr01 (accessed March 4, 2014).
16. Donaldson et al., *Heat and Cold*, 136.
17. Texas State Historical Association, "Refrigeration," http://www.tshaonline.org/handbook/online/articles/dqr01 (accessed March 4, 2014).
18. Wikipedia, "Ice Trade," http://en.wikipedia.org/wiki/Ice_trade (accessed March 4, 2014).
19. Gantz, *Vacuum Cleaner*.
20. Cheney, *Tesla*, 32–33.
21. Wikipedia, "War of Currents," http://en.wikipedia.org/wiki/War_of_Currents (accessed March 4, 2014).
22. Wikipedia, "Carl von Linde," http://en.wikipedia.org/wiki/Carl_von_Linde (accessed March 4, 2014).
23. Donaldson et al., *Heat and Cold*, 131.
24. Ibid., 132.
25. Burke, *Connections*, 244.
26. Wikipedia, "Carl von Linde," http://en.wikipedia.org/wiki/Carl_von_Linde (accessed March 4, 2014).
27. Donaldson et al., *Heat and Cold*, 136.
28. Ibid., 138.
29. Ibid., 139.
30. Ibid., 141.
31. Ibid., 139.
32. Ibid., 144.

33. Rees, *Refrigeration Nation.*
34. Donaldson et al., *Heat and Cold,* 141.
35. Ibid., 53.
36. Ibid., 147.
37. Ibid., 52–53.
38. Texas State Historical Association, "Refrigeration," http://www.tshaonline.org/handbook/online/articles/dqr01 (accessed May 27, 2014).
39. Donaldson et al., *Heat and Cold,* 52–53.
40. Cowan, *More Work for Mother,* 130.
41. Donaldson et al., *Heat and Cold,* 52.
42. Wikipedia, "Gibson Appliance," http://en.wikipedia.org/wiki/Gibson_Appliance (accessed May 27, 2014).
43. Donaldson et al., *Heat and Cold,* 51.
44. Ibid., 57.
45. Cowan, *More Work for Mother,* 131.
46. Wikipedia, "Icebox," http://en.wikipedia.org/wiki/Icebox (accessed May 27, 2014).
47. Wikipedia, "James Dewar," http://en.wikipedia.org/wiki/James_Dewar (accessed May 27, 2014).
48. Donaldson et al., *Heat and Cold,* 145.
49. Ibid., 181.

Chapter 4

1. Donaldson et al., *Heat and Cold,* 191.
2. Nagengast, "A History of Comfort Cooling Using Ice," 50.
3. Donaldson et al., *Heat and Cold,* 269.
4. Nagengast, "A History of Comfort Cooling Using Ice," 51.
5. Ibid., 52–53.
6. Ibid., 50.
7. Donaldson et al., *Heat and Cold,* 161.
8. Nagengast, "A History of Comfort Cooling Using Ice," 53.
9. Donaldson et al., *Heat and Cold,* 268–69.
10. Nagengast, "A History of Comfort Cooling Using Ice," 53.
11. "History of Refrigeration Timeline: 1906–1939," https://www.ashrae.org/resources—publications/history-of-refrigeration-timeline—1906–1939 (accessed May 27, 2014).
12. Dolin, "Refrigeration Past, Present and Future."
13. Donaldson et al., *Heat and Cold,* 331, ch. 11, note 56.
14. Ibid., 273.
15. Donaldson et al., *Heat and Cold,* 331, ch. 11, note 43.
16. Ibid., 274.
17. Ibid., 275.
18. "Air Conditioning and Refrigeration History: Part 3," http://www.greatachievements.org/?id=3862 (accessed May 27, 2014).
19. Donaldson et al., *Heat and Cold,* 277.
20. Wikipedia, "Willis Carrier," http://en.wikipedia.org/wiki/Willis_Carrier (accessed May 27, 2014).
21. Donaldson et al., *Heat and Cold,* 281.
22. Ibid., 283.
23. "A Brief History of Air Conditioning," http://www.popularmechanics.com/home/improvement/electrical-plumbing/a-brief-history-of-air-conditioning-10720229?click=main_sr (accessed June 22, 2014).
24. Reyner Banham as quoted in Donaldson et al., *Heat and Cold,* 329, ch. 11, note 2.
25. Donaldson et al., *Heat and Cold,* 331, ch. 11, note 56.
26. "Air Conditioning and Refrigeration History: Part 2," http://www.greatachievements.org/?id=3859 (accessed June 22, 2014).
27. Donaldson et al., *Heat and Cold,* 287.
28. Ibid., 284.
29. Ibid., 333, ch. 11, note 88.
30. Ibid.
31. Ibid., note 89.
32. Ibid., 286.
33. Ibid., 288.
34. Cowan, *More Work for Mother,* 131.
35. Wikipedia, "Ice Trade," http://en.wikipedia.org/wiki/Ice_trade (accessed June 22, 2014).
36. Donaldson et al., *Heat and Cold,* 209.
37. Ibid., 205–06.
38. Ibid., 207.
39. Cowan, *More Work for Mother,* 133.
40. Donaldson et al., *Heat and Cold,* 207.

41. Ibid., 206.
42. Ibid., 223.
43. Ibid., 221–22.
44. Ibid., 324, ch. 9, note 97.
45. Ibid., 208–09.
46. Ibid., 327, ch. 10, note 128.
47. Ibid., 208–09.
48. Ibid., 207.
49. Cowan, *More Work for Mother*, 130.
50. Grand Rapids Historical Commission, "Grand Rapids Refrigerator Co.," http://www.furniturecityhistory.org/company/4666/grand-rapids-refrigerator-co (accessed June 22, 2014).
51. Wikipedia, "Leonard (Appliances)," http://en.wikipedia.org/wiki/Leonard_(appliances) (accessed June 22, 2014).
52. Cowan, *More Work for Mother*, 132.
53. Donaldson et al., *Heat and Cold*, 211.
54. Cowan, *More Work for Mother*, 132.
55. Donaldson et al., *Heat and Cold*, 211.
56. Ibid., 211–14.
57. Patten, *Made in USA*, 244.
58. Donaldson et al., *Heat and Cold*, 214.
59. Patten, *Made in USA*, 242.
60. Donaldson et al., *Heat and Cold*, 214.
61. Patten, *Made in USA*, 245.
62. Donaldson et al., *Heat and Cold*, 214.
63. Patten, *Made in USA*, 244.
64. Donaldson et al., *Heat and Cold*, 214.
65. Ibid., 238.
66. Ibid., 210–11.
67. Ibid., 211.

Chapter 5

1. Donaldson et al., *Heat and Cold*, 220.
2. Ibid., 220–21.
3. Rees, *Refrigeration Nation.*
4. Nickels, "Preserving Women," 696.
5. Gantz, *The Vacuum Cleaner*, 93.
6. Donaldson et al., *Heat and Cold*, 223–25.
7. Rees, *Refrigeration Nation.*
8. Cowan, *More Work for Mother*, 133.
9. Ibid.
10. Ibid., 141.
11. Ibid., 140–42.
12. Wikipedia, "Einstein Refrigerator," http://en.wikipedia.org/wiki/Einstein_refrigerator (accessed June 22, 2014).
13. Donaldson et al., *Heat and Cold*, 222–25.
14. Ibid., 225.
15. Hall of History, "*General Electric Story*," 12.
16. Rees, *Refrigeration Nation.*
17. Ibid.
18. Donaldson et al., *Heat and Cold*, 225.
19. Friday, "*Walk Through the Park*," 16.
20. Donaldson et al., *Heat and Cold*, 226.
21. Miksinski (1970) quoted in Rees, *Refrigeration Nation.*
22. Cowan, "How the Refrigerator Got Its Hum."
23. Donaldson et al., *Heat and Cold*, 226.
24. Ibid., 228.
25. Ibid., 325, ch. 10, note 118.
26. Ibid., 228.
27. Rees, *Refrigeration Nation.*
28. Wikipedia, "Dry Ice," http://en.wikipedia.org/wiki/Dry_ice (accessed June 22, 2014).
29. Wikipedia, "Icyball," http://en.wikipedia.org/wiki/Icyball (accessed June 22, 2014).
30. "A Brief History of Air Conditioning," http://www.popularmechanics.com/home/improvement/electrical-plumbing/a-brief-history-of-air-conditioning-10720229?click=main_sr (accessed June 22, 2014).
31. Donaldson et al., *Heat and Cold*, 290.
32. Ibid.
33. Ibid., 292.
34. Ibid., 294.
35. *Scientific American*, January 1930, 14–17.
36. Donaldson et al., *Heat and Cold*, 295.
37. Ibid., 295–301.
38. Ibid., 179.
39. Christine Frederick as quoted in Gorman, *Industrial Design Reader*, 92–96.
40. Christine Frederick as quoted in Avakian et al., *From Betty Crocker to Feminist Food Studies*, 50.
41. Cowan, *More Work for Mother*, 175.
42. Ibid., 177.

43. Ibid., 175.
44. Donaldson et al., *Heat and Cold*, 242.

Chapter 6

1. Lifshey, *Housewares Story*, 131.
2. Gantz, *Design Chronicles*, 59.
3. Redell Story as quoted in Lifshey, *Housewares Story*, 139.
4. Ibid., 138–39.
5. Ibid., 142–44.
6. Pulos, *American Design Ethic*, 304.
7. Helen Read as quoted in Gorman, *Industrial Design Reader*, 116.
8. Pulos, *American Design Ethic*, 332.
9. Elmo Calkins as quoted in Pulos, *American Design Ethic*, 316.
10. Elmo Calkins as quoted in Pulos, *American Design Ethic*, 324.
11. Patten, *Made in USA*, 235.
12. Pulos, *American Design Ethic*, 332.
13. Elmo Calkins as quoted in Gorman, *Industrial Design Reader*, 130.
14. Elmo Calkins as quoted in Gorman, *Industrial Design Reader*, 131.
15. Franklin D. Roosevelt, inaugural address, March 4, 1933.
16. Donaldson et al., *Heat and Cold*, 301.
17. Ibid., 301–02.
18. Ibid., 302.
19. Ibid., 302–07.
20. Ibid., 307.
21. "Air Conditioning and Refrigeration Timeline," http://www.greatachievements.org/?id=3854 (accessed June 22, 2014).
22. "A Brief History of Air Conditioning," http://www.popularmechanics.com/home/improvement/electrical-plumbing/a-brief-history-of-air-conditioning-10720229?click=main_sr (accessed June 22, 2014).
23. Wikipedia, "Gibson Appliance," http://en.wikipedia.org/wiki/Gibson_Appliance (accessed June 22, 2014).
24. Rees, *Refrigeration Nation*.
25. Donaldson et al., *Heat and Cold*, 326, ch. 10, note 123.
26. Friday, *Walk Through the Park*, 16–17.

Chapter 7

1. Van Doren, "Streamlining."
2. Wikipedia, "Powel Crosley, Jr.," http://en.wikipedia.org/wiki/Powel_Crosley,_Jr. (accessed June 22, 2014).
3. Patten, *Made in USA*, 245.
4. Wikipedia, "Powel Crosley, Jr.," http://en.wikipedia.org/wiki/Powel_Crosley,_Jr. (accessed June 22, 2014).
5. Oliver Sprague as quoted in Nelson, "Both Fish and Fowl."
6. Patten, *Made in USA*, 247.
7. Ibid.
8. Loewy, *Industrial Design*, 100.
9. Patten, *Made in USA*, 247.
10. Loewy, *Industrial Design*, 98.
11. Cowan, "How the Refrigerator Got Its Hum."
12. Loewy, *Industrial Design*, 100.
13. Ibid.
14. Patten, *Made in USA*, 247.
15. Nickles, "Preserving Women."
16. Ibid.
17. Lifshey, *Housewares Story*, 132.
18. Wikipedia, "Evaporative Cooler," http://en.wikipedia.org/wiki/Evaporative_cooler (accessed June 22, 2014).
19. 1939 World's Fair Guidebook.
20. Friday, *Walk Through the Park*, 17.
21. Nagengast, "A History of Comfort Cooling Using Ice," 53–56.
22. Wikipedia, "Clarence Birdseye," http://en.wikipedia.org/wiki/Clarence_Birdseye (accessed June 22, 2014).
23. Donaldson et al., *Heat and Cold*, 312, ch. 4, note 13.
24. Ibid., 333, ch. 11, note 75.

Chapter 8

1. Funding Universe, "Fedders Corp. History," http://www.fundinguniverse.com/company-histories/fedders-corp-history/ (accessed June 22, 2014).
2. Fedders, "About Fedders: 116 Years of Evolution," http://www.fedders.com/history.html (accessed June 22, 2014).
3. Muller-Munk, "Industrial Designers," 116.
4. Dreyfuss, "Silent Salesman of Industry," 2.

5. "Servel, Inc.: A Case Study in Product Change and a Search for New Markets."
6. Jimmy Scichilone, collector, personal correspondence with author.
7. Wikipedia, "Automobile Air Conditioning," http://en.wikipedia.org/wiki/Automobile_air_conditioning (accessed June 22, 2014).
8. Funding Universe, "Fedders Corp. History," http://www.fundinguniverse.com/company-histories/fedders-corp-history/ (accessed May 12, 2014).
9. Bertin, "House and Garden Color Program."
10. Funding Universe, "Fedders Corp. History," http://www.fundinguniverse.com/company-histories/fedders-corp-history/ (accessed June 22, 2014).
11. Wikipedia, "Gibson Appliance," http://en.wikipedia.org/wiki/Gibson_Appliance (accessed June 22, 2014).
12. Wikipedia, "Cryogenics," http://en.wikipedia.org/wiki/Cryogenics (accessed June 22, 2014).

Chapter 9

1. Wikipedia, "Magnetic Refrigeration," http://en.wikipedia.org/wiki/Magnetic_refrigeration (accessed June 3, 2014).
2. Dolin, "Refrigeration Past, Present and Future."
3. Rizzo, "New Nuclear Freeze."
4. *The Weekly Standard* magazine, May 5, 2014.

Chapter 10

1. Wikipedia, "Snowmaking," http://en.wikipedia.org/wiki/Snowmaking (accessed June 3, 2014).

Bibliography

Adamson, Glenn. *Industrial Strength Design*. Cambridge, MA: MIT Press, 2003.

Amato, Ivan. *Stuff*. New York: Basic, 1997.

Avakian, Arlene Voski, and Barbara Haber. *From Betty Crocker to Feminist Food Studies*. UK: Liverpool University Press, 2005. http://scholarworks.umass.edu/cgi/viewcontent.cgi?article=1000&context=umpress_fbc. Accessed May 7, 2014.

Beecher, Catherine, and Harriet Beecher Stowe. *The American Woman's Home; or, Principles of Domestic Science*. New York: J.B. Ford, 1869. http://www.gutenberg.org/cache/epub/6598/pg6598.html. Accessed May 07, 2014.

Bernstein, Mark. "Thomas Midgely and the Law of Unintended Consquences." *Invention and Technology* 17, no. 4 (Spring 2002): 38–46.

Bertin, Nadine. "The House and Garden Color Program." *Color Research and Application* 3, no. 2 (Summer 1978): n.p.

Bunch, Bryan, and Alexander Hellemans. *The Timetables of Technology*. New York: Touchstone, 1993.

Burke, James. *Connections*. New York: Little, Brown, 1978.

Cheney, Margaret. *Tesla: Man Out of Time*. New York: Barnes and Noble, 1993.

Cowan, Ruth Schwartz. "How the Refrigerator Got Its Hum." In *The Social Shaping of Technology*, ed. Donald A. MacKensie and Judy Wajcman, 202–218. Philadelphia : Open University Press, 1985.

_____. *More Work for Mother*. New York: Basic, 1983.

Derry, T.K., and Trevor I. Williams. *A Short History of Technology*. New York: Dover, 1960.

Dolin, Brian. "Refrigeration Past, Present and Future." KE2 Therm Solutions, Washington, Missouri, 2012. http://www.ke2therm.com/files/W-9-5_Refrig_Past_and_Future_Jan_12.pdf. Accessed May 7, 2014.

Donaldson, Barry, Bernard Nagengast, and Gershon Mekler. *Heat and Cold: Mastering the Great Indoors*. Atlanta, GA: American Society of Heating, Refrigerating, and Air Conditioning Engineers, Inc., 1994.

Dreyfuss, Henry. "The Silent Salesman of Industry." Dinner address, May 28, 1952, at the Industrial Design Conference, 81st annual meeting of the Canadian Manufacturing Association, Toronto, Canada, reprinted in *Industrial Canada*, July 1952, 2.

Eales, Mrs. Mary. *Mrs. Mary Eales's Receipts*. London: Printed for J. Robson. 1733. http://www.gutenberg.org/files/20735/20735-h/20735-h.htm. Recipe #93. Accessed May 7, 2014.

Friday, Franklin. *A Walk Through the Park: The History of GE Appliances and Appliance Park.* Louisville, KY: Elfun Historical Society, 1987.

Gantz, Carroll. *Design Chronicles.* Atglen, PA: Schiffer, 2005.

_____. *The Industrialization of Design.* Jefferson, NC: McFarland, 2010.

_____. *The Vacuum Cleaner.* Jefferson, NC: McFarland, 2011.

Gorman, Carla R. *The Industrial Design Reader.* New York: Allworth, 2003.

Ikenson, Ben. *Patents: Ingenious Inventions.* New York, Black Dog and Leventhal, 2004.

Industrial Designers Society of America. *Design in America.* New York: McGraw Hill, 1969.

Hall of History. *The General Electric Story: 1876–1986.* Schenectady, NY: Hall of History Foundation, 1989.

James, Peter, and Nick Thorpe. *Ancient Inventions.* New York: Ballantine, 1994.

Lifshey, Earl. *The Housewares Story.* Chicago: National Housewares Manufacturers Association, 1973.

Loewy, Raymond. *Industrial Design.* Woodstock, NY: Overlook, 1979.

Macauley, David. *The Way Things Work.* Boston: Houghton Mifflin, 1988.

Matranga, Victoria. Personal correspondence with author.

McDermott, Catherine. *Twentieth Century Design.* Woodstock, NY: Overlook, Peter Mayer, 2000.

Messadié, Gerald, *Great Modern Inventions.* Edinburgh, W.R. Chambers, 1991.

Miksinski, Anne Francis. *The Quiet Revolution: The History and Effects of Domestic Refrigeration in America, from Ice to Mechanical Refrigerants.* M.A thesis, George Washington University, 1970.

Muller-Munk, Peter. "Industrial Designers Can Face Today's Challenge." *Electrical Manufacturing,* June 1951, 116.

Nagengast, Bernard. "A History of Comfort Cooling Using Ice." *ASHRAE Journal,* February 1999, 49–57. https://www.ashrae.org/.../docLib/.../200362710047_326.pdf. Accessed May 7, 2014.

Nelson, George. "Both Fish and Fowl." *Fortune,* February 1934, 40–98.

Nickles, Shelley. "Preserving Women: Refrigerator Design as Social Progress in the 1930s." *Technology and Culture* 43, no. 4 (October 2002): 693–727. http://muse.jhu.edu/login?auth=0&type=summary&url=/journals/technology_and_culture/v043/43.4nickles.pdf. Accessed May 7, 2014.

Norman, Donald A. *The Design of Everyday Things.* New York: Doubleday, 1988.

Patton, Phil. *Made in USA.* UK: Penguin, 1992.

Pulos, Arthur. *The American Design Adventure.* Cambridge, MA: MIT Press, 1988.

_____. *American Design Ethic.* Cambridge, MA: MIT Press, 1983.

Raffald, Elizabeth. *The Experienced English Housekeeper.* Manchester, UK: Printed by J. Harrop, for the author, 1769.

Rees, Jonathon. *Refrigeration Nation: A History of Ice, Appliances, and Enterprise in America.* Baltimore, MD: John Hopkins University Press, 2013. http://books.google.com/books?id=JeoEAQAAQBAJ&pg=PT16&lpg =PT16&dq=refrigeration+nation&source=bl&ots=BcQM5UOv1T&si g=hcx4XI2LJBO6mNGHuHndJgBgqbw&hl=en&sa=X&ei=UtlnU5 MGqfDJAeWvgMAO&ved=0CEoQ6AEwAzgK#v=onepage&q=refri geration%20nation&f=false. Accessed May 7, 2014.

Rizzo, Johnna. "A New Nuclear Freeze." *National Geographic,* March 2014, 27.

Scichilone, Jimmy. Personal correspondence with author.

Scientific American, January 1930, 14–17.

"Servel, Inc.: A Case Study in Product Change and a Search for New Markets." *Tide,* January 16, 1953.

Society of Industrial Designers. *Industrial Design in America 1954.* New York: Farrar, Straus and Young, 1954.

Storey, Redell. *New York Times,* August 1927.

Stradley, Linda. "Ice Cream History." What's Cooking America. Ice Cream-History, legends, and Myths of Ices and Ice Cream. 2004. http://whatscookingamerica.net/History/IceCream/IceCreamHistory.htm. Accessed May 7, 2014.

Van Doren, Harold. "Streamlining: Fad or Function." *Design* [UK] 10, October 1949, 2–28.

Van Dulken, Stephen. *Inventing the 20th Century.* New York University Press, 2000.

Wayt, Hampton. Personal correspondence with author.

Index

Numbers in ***bold italics*** indicate pages with photographs.